The Master Molecule:
Abscisic Acid and
Plant Resilience in Water Stress

Salina

INDEX

INTRODUCTION

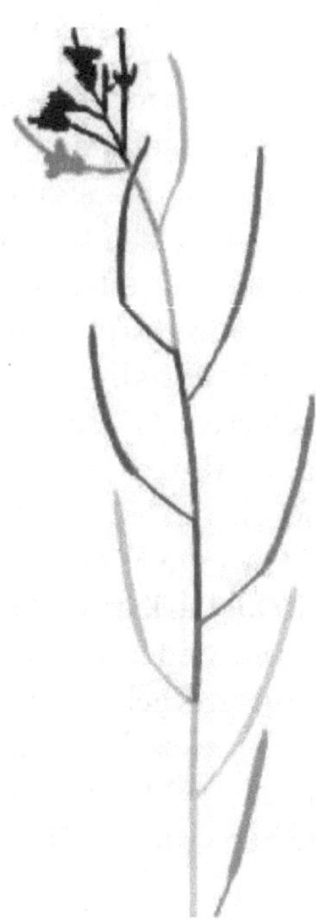

ABA plays a pivotal role to coordinate plant response under water stress situations as well as to regulate plant growth and development (Cutler et al., 2010). Chemically, ABA ($C_{15}H_{20}O_4$) is a sesquiterpenoid derived from isopentenyl pyrophosphate synthesized in plastids through the 2C-methyl-D-erythritol4-phosphate (MEP) pathway (Nambara & Marion-Poll, 2005). ABA contains one asymmetric carbon atom at C1′, the natural form is S(+)ABA and the side chain of the molecule is present in the 2-cis,4-trans isomeric state. A plane of symmetry can be defined in the ABA molecule through the optical centre, which defines two sides that differ only in that C6′ carries two methyl group whereas C2′ carries one and a double bond. Consequently, the non-natural (-) enantiomer only differs structurally in these positions from the (+) enantiomer (Milborrow, 1974).

Plant hormone research has greatly benefited from genetic screenings aimed at the identification of key components of the hormone signaling pathways. However, such screenings failed to identify ABA receptors. Functional redundancy or pleiotropic effects including embryo or gamete lethality looked as sound arguments to justify this failure (Santiago et al., 2012). Finally, a chemical genetic approach using a synthetic selective ABA agonist, pyrabactin, made it possible to identify a family of soluble ABA receptors, named PYRABACTIN RESISTANCE1 (PYR1) / PYR1-LIKE (PYL) / REGULATORY COMPONENTS OF ABA RECEPTORS (RCAR) (Ma et al., 2009; Park et al., 2009). Mutant plants lacking these receptors presented indeed ABA-insensitive phenotypes, as the sextuple mutant lacking PYR1, PYL1, PYL2, PYL4, PYL5 and PYL8 (Gonzalez-Guzman et al., 2012), or the duodecuple mutant *pyr1 pyl1 pyl2 pyl3 pyl4 pyl5 pyl7 pyl8 pyl9 pyl10 pyl11 pyl12* (Zhao et al., 2018).

PYR/PYL/RCAR ABA receptors.

PYR/PYL/RCAR are part of the superfamily START/Bet v proteins, which are characterized by the presence of a cavity where hydrophobic ligands can be incorporated (Radauer et al., 2008). In the case of the PYR/PYL/RCAR family, this cavity is the ABA-binding pocket, where the hormone establishes numerous hydrophobic and polar interactions (Santiago et al., 2012; Umezawa et al., 2009). This family is composed of 14 members in *Arabidopsis thaliana*, which are also present as multiple families in crop species ranging from 15 putative members in tomato (Gonzalez-Guzman et al., 2014), 12 in rice (He et al., 2014) or 13 in maize (He et al., 2018). They are distributed in distinct subfamilies in *A. thaliana*. PYR1 and PYL1-PYL3 constitute a subfamily of dimeric receptors (clade III), while PYL4-PYL6 (clade II) and PYL7-PYL10 (clade I) subfamilies are monomeric receptors (Dupeux et al., 2011b). PYL11-PYL13 oligomeric state has not been analysed, likely because of their poor solubility (Li et al., 2013).

Not all receptors contribute equally to the ABA response. This can be figured out by genetic analysis, as for example: the pentuple mutant *pyl3 pyl7 pyl9 pyl11 pyll12* shows wild type sensitivity to ABA, whereas the quadruple mutant *pyr1 pyl1 pyl2 pyl4* show a clear ABA-insensitive phenotype (Gonzalez-Guzman et al., 2012; Park et al., 2009; Zhao et al., 2018). Regarding their biochemical properties, dimeric receptors have lower intrinsic affinity for ABA than the monomeric receptors. Therefore, at basal ABA levels the dimeric receptors might be less active than monomeric receptors. However, in the presence of clade A protein phosphatases type 2C (PP2C) co-receptor, both dimeric and monomeric receptors show nanomolar affinity for ABA. Dimeric receptors require dimer dissociation during the activation process, because the dimerization interface overlaps with the PP2C binding region (Dupeux et al., 2011b). As a result, overexpression of monomeric ABA receptors led to an increased water use efficiency (WUE) in contrast to dimeric receptors (Tischer et al., 2017; Yang et al., 2019).

The study of single receptors in different biological contexts has revealed specific roles. The *pyl8* single mutant reported a non-redundant role in root sensitivity to ABA and regulation of lateral root growth (Antoni et al., 2013; Zhao et al., 2014). PYL8 perception of ABA in root tissue involves a non-cell-autonomous mechanism and is also a unique receptor showing ABA-induced stabilization and predominant nuclear localization (Belda-Palazon et al., 2018). In contrast, other receptors show cytoplasmic and nuclear localization. Secondly, the *pyl9* single mutant showed reduced ABA-induced leaf senescence under low light (Zhao et al., 2016). ABA signaling is also involved in the induction of seed and bud dormancy under unfavourable conditions, and specific roles for other ABA receptors might be reported in this process (Gonzalez-Grandio et al., 2017). Additionally, specific receptors play different roles in the regulation of stomatal closure, PYL2 is sufficient for guard cell ABA-induced responses, whereas in responses to CO_2, PYL4 and PYL5 are essential (Dittrich et al., 2019). The *pyr1 pyl1 pyl2 pyl4* quadruple mutant was severely impaired both in ABA-induced stomatal closure and ABA-inhibition of stomatal opening after light exposure (Nishimura et al., 2010). Finally, phaseic acid (PA), an ABA catabolite, is able to activate some ABA receptors, such as PYL3 (Weng et al., 2016). This suggests that functional diversity of the receptors has evolved upon expansion of the gene family.

Certain functional redundancy occurs because to obtain severe ABA-insensitive phenotype, inactivation of six ABA receptors is required (Gonzalez-Guzman et al., 2012) and several ABA receptors are needed in roots to mediated adaptive responses to water deficit such as hydrotropism and repression of lateral root formation (Antoni et al., 2013; Dietrich et al., 2017; Orman-Ligeza et al., 2018). In rice (*Oryza sativa*), both redundant and differential functions have been observed. Clade I mutants showed seed dormancy and stomatal movement defects, in contrast to clade II mutants, that did not. Also, the *pyl1 pyl4 pyl6* triple mutant showed improved growth and productivity (Miao et al., 2018), likely because of reduced ABA mediated growth inhibition under non-stress conditions.

The ABA Core Signaling Pathway.

The ABA signaling pathway (**Figure 1**) is initiated by ABA perception through PYR/PYL/RCAR receptors. However, dimeric receptors have low intrinsic affinity for ABA in the absence of the PP2C co-receptor (Kd over 50µM) and likely exist as inactive homodimers in cells, unable to bind or inhibit PP2Cs at basal ABA levels. However, in the presence of submicromolar ABA and the PP2C, they achieve nanomolar affinity for ABA binding (Santiago et al., 2009). Monomeric receptors show higher intrinsic affinity for ABA in the absence of the PP2C (Kd circa 1µM), basically because dimeric receptors suffer the thermodynamic penalty imposed by dimer dissociation during the receptor activation process (Dupeux et al., 2011b). In order to obtain nanomolar affinity, PP2Cs are required as co-receptors, leading to ternary receptor-ABA-phosphatase complexes (Moreno-Alvero et al., 2017). The first experiments that led to the discovery of this interaction between the phosphatases and the receptors was a yeast two-hybrid assay (Y2H) (Ma et al., 2009; Santiago et al., 2009). The binding of ABA to the receptors produces a structural reorganization that enables docking into the phosphatase active site to inhibit its activity (Soon et al., 2012). The structure of the complex revealed that the gate loop inserts into the active site of the phosphatase blocking its activity (Moreno-Alvero et al., 2017). Therefore, ABA receptors in the presence of ABA are very effective competitive inhibitors of

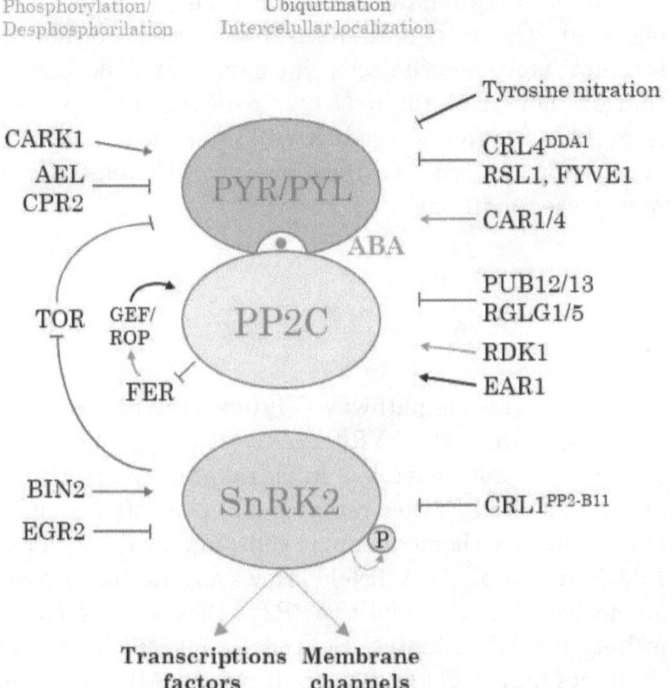

Figure 1. The ABA core signaling pathway is based on the reception of the ABA by the binary receptor complex (PYR/PYL-ABA-PP2C) releasing the protein kinases (SnRK2) from their inhibition, which then act as response mediators to the targets. The three core ABA signaling components are post-translationally regulated. PYR/PYLs are phosphorylated by AELs, CPR2, CARK1 and TOR kinase. TOR is phosphorylated by SnRK2s. The SnRK2s are phosphorylated and dephosphorylated by BIN2 and clade E PP2C EGR2, respectively. ABI2 PP2C dephosphorylate FER, the receptor of RALF, that activates ABI2 through the GEF–ROP pathway. The core components are ubiquitinated by CRL4DDA1 and RSL1 for PYR/PYLs, by PUB12/13, RGLG1/5 and PIR1/2 for PP2Cs, and by CRL1PP2-B11 for SnRK2s. Intracellular localization of PYR/PYLs and PP2Cs is controlled by a small lipid-binding protein (CAR) or a receptor-like kinase (RDK1), respectively. PYR/PYLs inhibition over PP2Cs is diminished partially or fully by Tyr nitration or S-nitrosylation, respectively. EAR1 is a protein of unknown function that appears to positive regulate the PP2Cs. Adapted from (Yoshida et al., 2019)

the physiological substrates of the PP2Cs (Melcher et al., 2009). These substrates include three ABA-activated SNF1-

related protein kinases (SnRK2s). Once the kinases are relieved from their inhibition, they can begin a cascade of phosphorylations producing the ABA signal response (Umezawa et al., 2009; Vlad et al., 2009).

Clade A protein phosphatases type 2C.

Phosphatases are enzymes that remove a phosphate group from a phosphorylated amino acid. This amino acid can be a serine (Ser), a threonine (Thr) or a tyrosine (Tyr), and the phosphatases can be classified be their substrate specificity. Type 1 phosphatases (PP1) prefers the beta-subunit of phosphorylase kinase as substrate, while type 2 phosphatases (PP2) preferentially dephosphorylate the alpha-subunit. PP2s can be further subdivided by their dependence on divalent cations into PP2A, PP2B and PP2C. Whereas PP2B and PP2C are regulated by Ca^{2+} and Mg^{2+}, respectively, PP2A, as PP1, does not require divalent cations for activity. A group of drugs have also been useful in distinguishing members of each type, for example, both okadaic acid and calyculin A potently inhibit the activity of PP1 and PP2A but are not effective to PP2B and PP2C (Luan, 2003). In *A. thaliana* there are 76 putative Ser/Thr PP2Cs proteins subdivided in 10 different clades, greatly outnumbering other eukaryotes (Schweighofer et al., 2004), as for example, the 15 found in human (Cheng et al., 2000). PP2C functions emphasize the existence of sophisticated signaling pathways in plants, in which protein dephosphorylation plays a crucial role towards determining specificities (Schweighofer et al., 2004). The clade A of this phosphatases are central negative regulators of ABA signaling in seeds and also in vegetative tissues (Antoni et al., 2012; Gosti et al., 1999). There are 9 members in the clade A of the PP2Cs in *A. thaliana*, which are subdivided in two separate branches based on their amino acid sequence alignments (Schweighofer et al., 2004). ABA INSENSITIVE 1

(ABI1) (the first PP2C identified as involved in ABA signaling), ABI2, HYPERSENSITIVE TO ABA 1 (HAB1) and HAB2 conform the first branch, while the members of the second branch are ABA HYPERSENSITIVE GERMINATION 1 (AHG1), PROTEIN PHOSPHATASE 2CA (PP2CA/AHG3), HIGHLY ABA INDUCIBLE 1 (HAI1), HAI2 and HAI3.

PP2Cs are localized in cytoplasm and nucleus, which agrees with their reported interaction with SnRK2s (Fujita et al., 2009). One important structural feature of these phosphatases is a conserved tryptophan (Trp) residue (Trp300 in ABI1 and Trp385 in HAB1). A mutation of this residue to alanine (Ala), blocks the union of ABA avoiding the formation of the ternary complex, essential for the ABA-dependant inactivation of the phosphatase by PYR/PYLs. However these mutants could still target the SnRK2s, making the plants ABA insensitive (Dupeux et al., 2011a). Another mutation of the phosphatases that leads to their escape from the receptors is a conserved glycine (Gly) residue (Gly180 in ABI1 and Gly246 in HAB1), which was mutated to aspartic acid (Asp) (Leung et al., 1994; Leung et al., 1997; Meyer et al., 1994; Rodriguez et al., 1998). The bulky Asp likely prevents the receptors from establishing contact, producing in these plants strong ABA insensitivity in seed and vegetative tissues (Santiago et al., 2012; Umezawa et al., 2009; Vlad et al., 2009).

These phosphatases are able to regulate subclass III SnRK2s by physically blocking their kinase active site and dephosphorylate the conserved Ser residue in their activation loop (Soon et al., 2012). But they also target important effectors of ABA signaling such as SLOW ANION CHANNEL 1 (SLAC1) and ABRE-binding transcription factors (Antoni et al., 2012; Brandt et al., 2015; Lee et al., 2009; Lynch et al., 2012). PP2CA inhibits SLAC1 by direct interaction with it and also by dephosphorylating the SnRK2, blocking the phosphorylation of SLAC1 (Lee et al., 2009). Furthermore, some of the PP2Cs play key roles to regulate seed germination, as for example AHG1, which plays an important role regulating ABA signaling in seeds. Additionally, AHG1 and PP2CA can regulate with the help of DELAY OF GERMINATION 1 (DOG1) seed dormancy and consequently the timing of seed germination (Nee et al., 2017; Nishimura et al., 2018; Nishimura et al., 2007). Another example is AKT2 K(+) transport, that is regulated by PP2CA allowing the control of K(+) transport during stress situations and, consequently, influences membrane polarization (Cherel et al., 2002). Finally, taking in consideration the number of targets of the PP2Cs, we can state they act as a regulatory hub for different abiotic stress responses (Cherel et al., 2002; Forster et al., 2019; Geiger et al., 2009; Guo et al., 2002; Himmelbach et al., 2002; Ohta et al., 2003; Peirats-Llobet et al., 2016; Rodrigues et al., 2013; Sheen, 1996; Wang et al., 2018b; Yang et al., 2006). In agreement with this notion, combined inactivation of clade A PP2Cs enhances drought tolerance (Saez et al., 2006), as well as the overexpression of ABA receptors that leads to PP2C inhibition.

SNF1-related protein kinases 2.

Kinases are enzymes that catalyse the transfer of phosphate groups from high-energy, phosphate-donating molecules to specific substrates. Among the numerous kinases of the *A. thaliana* genome, for ABA signaling we pay particular attention to SnF1 related kinases (SnRKs). There are 38 SnRKs in *A. thaliana* that can be organized by their sequence similarity in 3 groups (3 SnRK1, 10 SnRK2 and 25 SnRK3) (Hrabak et al., 2003). While SnRK1 and SnRK2 subfamilies are calcium-independent kinases, SnRK3s are calcium dependent because they interact with calcineurin B-like (CBL) calcium sensors and therefore are also known as CBL-interacting protein kinases (CIPKs). Ca2+ signaling and CIPKs are known to be involved in ABA signaling (Edel &

Kudla, 2016), although SnRK2s are the canonical kinases that play a master role in ABA signaling.

SnRK2s are subdivided in 3 classes: subclass 1 kinases are activated by osmotic stress but no by ABA, subclass 2 are activated also by osmotic stress and weakly by ABA and subclass 3 are the only ones with a strong activation by osmotic stress and ABA. The subclass 3 kinases are SnRK2.2/SnRK2D, SnRK2.3/SnRK2I and SnRK2.6/SnRK2E/OST1 (Boudsocq et al., 2004). OST1 is highly expressed in guard cells, and therefore the *ost1* mutant is impaired in ABA-mediated stomatal closure (Mustilli et al., 2002; Yoshida et al., 2002). Single *snrk2.2* and *snrk2.3* mutants showed similar ABA sensitivity as compared to the wild type. However, the *snrk2.2 snrk2.3* double mutant shows ABA-insensitive phenotype in germination, dormancy and seedling growth (Fujii et al., 2007). The above results suggest a certain specialization of the three kinases, being OST1 more important as a regulator of stomatal closure, and SnRK2.2 and SnRK2.3 acting in other ABA responses. However, the triple mutant shows a severe ABA-insensitive phenotype, establishing the subclass 3 SnRK2s as core positive regulators of the ABA pathway (Fujii & Zhu, 2009). In contrast, other SnRK2s have more specific roles controlling the osmotic stress response (Fujii et al., 2011). Interestingly, in the sextuple and duodecuple PYR/PYL mutant plants, which are severely ABA-insensitive, the ABA-independent osmotic stress-induced activation of SnRK2s was strongly enhanced (Zhao et al., 2018). This suggests the existence of this alternative route when ABA-dependent responses are abolished in severe ABA-deficient or ABA-insensitive mutants.

Subclass 3 SnRK2s have an activation loop that needs to be phosphorylated in Ser/Thr residues to be activated. In the absence of ABA, clade A PP2Cs dephosphorylate these residues. On the contrary, in presence of ABA and with the phosphatases being repressed by the receptors, OST1 can autophosphorylate certain Ser residues of its activation loop

(Fujii et al., 2009; Mustilli et al., 2002; Umezawa et al., 2009; Vlad et al., 2009). In contrast, recombinant SnRK2.2 and SnRK2.3 kinases do not autophosphorylate under in vitro conditions (Zhu et al., 2017). SnRK2.6 can efficiently achieve autophosphorylation on its own due to its well-structured activation loop, which is in closer proximity to the active site than the activation loop of SnRK2.2 and SnRK2.3 (Ng et al., 2011).

Numerous downstream substrates of the kinases have been identified, which include transcription factors, ion channels, NADPH oxidases and plasma membrane ATPases. ABRE binding factors (ABFs) are key transcription factors of the core ABA signaling pathway which are activated upon phosphorylation by subclass 3 SnRK2 kinases (Fujita et al., 2013). There are also important ion and water channels activated by these kinases, such as SLAC1 and R-type ion channel 1 (QUAC1), as well as aquaporins, which are involved in ABA-induced stomatal closure (Geiger et al., 2009; Grondin et al., 2015; Imes et al., 2013). OST1 can also phosphorylate NADPH oxidases to produce reactive oxygen species (ROS) in guard cells (Sirichandra et al., 2010). The above targets of SnRK2s are positively regulated by phosphorylation, whereas the phosphorylation can also inhibit a protein function. For example, the POTASSIUM CHANNEL IN ARABIDOPSIS THALIANA 1 (KAT1), which impairs K^+ influx in guard cells and facilitates stomatal closure, is inhibited by OST1 (Sato et al., 2009a). Furthermore, plasma membrane H^+-ATPases can also be negatively regulated by phosphorylation, which leads to plasma membrane depolarization (Planes et al., 2015).

Transcription factors involved in ABA response.

Transcription factors (TFs) are proteins that bind to specific sequences of DNA to regulate gene expression. Genes that are ABA-inducible are identified by the presence of ABA responsive elements (ABREs) in their promoter sequence, and the transcription factor that recognize them are the ABRE-binding proteins (AREBs, ABFs) (Choi et al., 2000; Uno et al., 2000). ABFs form part of a bZIP transcription factor family (**Figure 2**). There are 9 members of this family in *A. thaliana*, from which ABF2, ABF3 and ABF4 are the most highly inducible by ABA and osmotic stress in vegetative tissues (Choi et al., 2000; Fujita et al., 2005; Uno et al., 2000; Yoshida

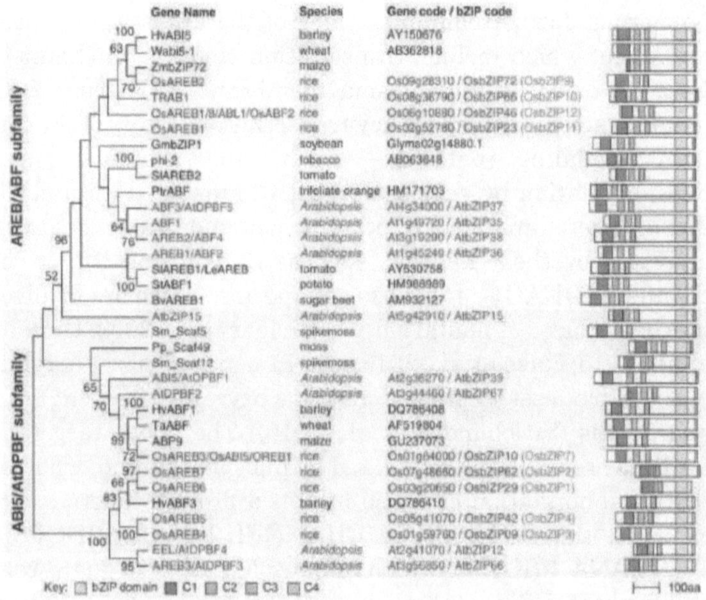

Figure 2. Phylogenetic tree and domain structure of bZIP TFs family and AREB/ABF and ABI5/AtDPBF subfamilies. bZIP TFs in *A. thaliana*, rice, bryophytes and other species of vascular plants are shown in red, green, blue and black letters, respectively. Reprinted from (Fujita et al., 2013).

et al., 2010). Experiments with the quadruple *abf1 abf2 abf3 abf4* mutant plants revealed these TFs control most of the downstream gene response to ABA (Yoshida et al., 2015).

ABI5 is also an important bZIP transcription factor from this family, which names its own subfamily (Lopez-Molina &

Chua, 2000). Experiments with the single *abi5* mutant revealed its key role as a positive regulator of ABA signaling during seed development, maturation, germination and early seedling growth (Lopez-Molina & Chua, 2000; Lopez-Molina et al., 2001; Lopez-Molina et al., 2002). Moreover, the expression of ABI5 is severely reduced in the *snrk2.2 snrk2.3 ost1* triple mutant plants, which suggests a regulation similar to the ABFs (Nakashima et al., 2009).

Besides ABREs, there are also other regulatory regions recognized by other ABA-related transcription factors. The drought response elements (DRE) are recognized by the APETALA2 (AP2) family of transcription factors, which includes ABI4 (Finkelstein et al., 1998). The single *abi4* mutant plants are insensitive to ABA (Soderman et al., 2000). ABI4 is a versatile transcription factor with many targets, regulating glucose response (Arenas-Huertero et al., 2000), nitrate-modulated root branching (Signora et al., 2001) and chloroplast-to-nucleus (Koussevitzky et al., 2007) and mitochondria-to-nucleus (Giraud et al., 2009) retrograde signaling pathways. Furthermore, ABI4 integrates redox, sucrose, ABA and jasmonate (JA) signaling (Kerchev et al., 2011).

ABI3 is another key transcription factor involved in ABA signaling. It belongs to the B3 family and recognizes ABRE regions in an ABA-dependant manner and RY/Sph regions in an independent way (Ezcurra et al., 2000). It has been reported that ABI3, ABI4 and ABI5 are co-expressed during seed maturation and seed germination. However, their levels decreased after this stages unless there is an external stress (Finkelstein et al., 2011).

Other transcription factors involved are WRKY40 (Shang et al., 2010) and WRKY63 (Ren et al., 2010) from the WRKY family that recognize W-box sequences, or MYC2 and MYB2, bHLH-related and MYB-related TFs that recognizes MYC or MYB regions respectively (Abe et al., 2003). Additional TFs that play crucial roles in response to ABA signaling in plants are NAC TFs (He et al., 2019) as SNAC-1, which is involved in drought response in *O. sativa* (Li et al., 2019a), or NF-Y TFs (Wang et al., 2019a), as NF-YC9, which interacts with ABI5 and facilitates its function in *A. thaliana* (Bi et al., 2017).

Regulation of ABA core components by post-transcriptional mechanisms and interacting proteins.

ABA signaling can be elicited by constitutively active PYR/PYL receptors, but a posttranscriptional mechanism limits their accumulation, suggesting a proteolytic mechanism to reduce their protein levels (Mosquna et al., 2011). The regulation of protein turnover is achieved by the ubiquitin-proteasome system or by the non-26S proteasome endomembrane trafficking pathway, and both are involved in the regulation of ABA signaling (Yu & Xie, 2017). The first proof of the ubiquitination of the receptors was achieved by p62-mediated affinity purification of ubiquitinated proteins in plant extracts treated with MG-132, a proteasome inhibitor. These plants were overexpressing HA-PYR1, HA-PYL4 and HA-PYL8, which then were detected in their ubiquitinated forms (Bueso et al., 2014a). The first described E3 ubiquitin ligases that target ABA receptors were CULLIN4-RING E3 ubiquitin ligases (CRL4) CRL4[DDA1] and RBR-type RING FINGER OF SEED LONGEVITY (RSL1) (Bueso et al., 2014b; Irigoyen et al., 2014). In the case of multimeric E3 ligases, a substrate adaptor is needed to recognize the target. In the case of the CRL4[DDA1] complex the substrate adaptor is DDB1-ASSOCIATED1 (DDA1), which can interact with PYL4, PYL5 and PYL8 promoting their degradation. However, in presence of ABA CRL4[DDA1] cannot promote PYL8 degradation (Irigoyen et al., 2014).

In addition to the ubiquitin-proteasome system, recent evidence revealed a dynamic turnover of ABA receptors from the plasma membrane through the non-26S proteasome endomembrane trafficking pathway (Belda-Palazon et al., 2016; Yu et al., 2016). This pathway is also triggered by ubiquitination of the target at the plasma membrane. Then, it requires endosomal trafficking and sorting of ubiquitinated proteins through the endosomal sorting complex required for transport (ESCRT) machinery (Yu & Xie, 2017). The internalization of PYR1 and PYL4 is initiated by the RBR-type E3 ubiquitin ligase RSL1 (Bueso et al., 2014b), and subsequently ubiquitinated ABA receptors are recognized by FYVE DOMAIN-CONTAINING PROTEIN 1 (FYVE1), also known as FYVE DOMAIN PROTEIN REQUIRED FOR ENDOSOMAL SORTING 1 (FREE1). With the help of VACUOLAR PROTEIN SORTING 23A (VPS23) and other components of the ESCRT machinery, ABA receptors are led to vacuolar degradation (Belda-Palazon et al., 2016; Yu & Xie, 2017). Consequently, ABA receptors are targeted for degradation both at the plasma membrane (Garcia-Leon et al., 2019) and the nucleus through RSL1 or CRL4[DDA1], respectively.

Additionally, there are others E3 ligases described as putative interactors of the ABA receptors, such as F-box E3 ligase RIFP1 that interacts with PYL8 (Li et al., 2016) or AtRAE1, which is a substrate adaptor of the multimeric CUL4-DDB1 E3 ligase, and is able to interact with PYL9 (Li et al., 2018). U-box E3 ubiquitin ligases PUB22 and PUB23 are also able to interact also with PYL9 (Zhao et al., 2017). However, endogenous levels of their targets have not been analysed. Finally, in *O. sativa*, The ANAPHASE PROMOTING COMPLEX/ CYCLOSOME (APC/C) is a multimeric E3 ligase that targets the OsPYL/RCAR receptors (Lin et al., 2015). This is interesting because it was supposed that most of the substrates targeted by APC/C are related to the control of the cell cycle. Tiller Enhancer (TE) is an activator of the APC/C complex, and through it the receptors are able to interact with the complex (Lin et al., 2015).

Ubiquitination is not the only post-translational mechanism that regulates the PYR/PYL ABA receptors. Mass spectrometry analysis of PYR/PYL receptors expressed in plant, identified additional post-translational modifications of the receptors (Castillo et al., 2015). Tyrosine nitration reduced the receptor inhibition of the phosphatases. The proteins that presented the tyrosine nitration were also polyubiquitinated, which suggest that probably the tyrosine nitration was enhancing the degradation of the receptors. Nitric oxide (NO)-deficient plants are ABA-hypersensitive, and tyrosine nitration is triggered by the combination of NO with superoxide ions, so this together suggest a rapid mechanism to inhibit ABA signaling by NO (Lozano-Juste & Leon, 2010). Moreover, S-nitrosylation of the receptors at cysteine residues was also revealed (Castillo et al., 2015).

Another fundamental post-translational mechanism for regulation is the phosphorylation and dephosphorylation. TARGET OF RAMPAMYCIN (TOR) complex is a kinase that phosphorylates the ABA receptors, inhibiting their interaction with PP2Cs. TOR phosphorylates a conserved Ser residue (Ser119 in PYL1) to prevent activation of the stress response. However, in presence of ABA, SnRK2s can phosphorylate Raptor, one component of the TOR complex, triggering the dissociation and inhibition of TOR (Wang et al., 2018b). ARABIDOPSIS EL1-LIKE (AEL) is a casein kinase that is also able to phosphorylate the ABA receptors. There are four described in *A. thaliana* (AEL1-AEL4). Triple *ael* mutant plants presented reduced phosphorylation, ubiquitination and degradation of PYR/PYLs, enhancing the ABA response. This suggest that the phosphorylation provided by AEL plays a role in the stability of the receptors, promoting their degradation (Chen et al., 2018). While AEL kinases phosphorylate the receptors in the nucleus, C-terminally encoded peptide receptor 2 (CEPR2) has a similar function in the plasma membrane (Yu et al., 2019). CEPR2 is a plasma membrane-localized Leucine-rich repeat receptor-like kinase (LRR-RLK) kinase able to interact there with the ABA receptors PYL2 and PYL4, and also with the C2-DOMAIN ABA-RELATED (CAR) proteins. CAR proteins do not modify the PYR/PYL receptors, but can change their localization positively regulating the ABA signaling. CARs mediate a transient Ca^{2+}-dependant interaction with phospholipid vesicles recruiting the receptors to the vesicles (Rodriguez et al., 2014). In absence of stress, CEPR2 might promote PYL ubiquitination in the plasma membrane for degradation, resulting in repressed ABA signaling. However, ABA inhibited the phosphorylation and degradation of PYL4 *in vitro* and *in vivo* (Yu et al., 2019). Finally, RECEPTOR-LIKE CYTOPLASMIC KINASE (RLCK), named in *A. thaliana* CARK1, is able to interact with PYR1, PYL1, PYL2, PYL3 and PYL8 phosphorylating them in one conserved Thr residue. Genetic analysis suggested that this phosphorylation is enhancing the receptors function, positively regulating the ABA signal. Plants overexpressing *CARK1*, presented indeed a higher drought resistance (Li et al., 2019b; Zhang et al., 2018b).

PYR/PYL receptors inhibit the activity of clade A PP2Cs by blocking their active site, but ABA receptors can also promote the degradation of the phosphatases. ABI1 is ubiquitinated by the plant U-box (PUB) E3 ligases PUB12 and PUB13, and ABA receptors and ABA are required to promote ABI1 ubiquitination (Kong et al., 2015). PUB12/13 cannot interact with other phosphatases, therefore, other E3 ligases are supposed to target the rest of clade A PP2Cs. RING DOMAIN LIGASE 1 (RGLG1) and RGLG5 RING-type E3 ligases were reported to target PP2CA for degradation, and this was enhanced by ABA (Wu et al., 2016). Additional RING ubiquitin E3 ligases, PP2CA interacting RING finger protein 1 (PIR1) and PIR2, interact with PP2CA and positively modulate ABA signaling by targeting PP2CA for degradation (Baek et al., 2019). Endogenous levels of PP2CA have not been analysed, however the *pir1 pir2* double mutant showed a clear ABA-insensitive phenotype.

Whereas PYR/PYLs inhibit PP2Cs, ENHANCER OF ABA CO-RECEPTOR 1 (EAR1) interacts with the PP2Cs to enhance their activity. EAR1 interacts with the N-terminal domain of the PP2Cs enhancing their activity and, therefore, negatively regulates the ABA signaling. Additionally, EAR1 accumulates in the nucleus in presence of ABA (Wang et al., 2018a). Furthermore, RECEPTOR DEAD KINASE 1 (RDK1) is a membrane protein, whose expression is induced by ABA. Once in the membrane, RDK1 can recruit ABI1 there, and this interaction was further enhanced by exogenous application of ABA. The *rdk1* single mutant showed diminished sensitivity to ABA-mediated inhibition of seedling establishment and root growth compared to Col-0 wild type, whereas RDK1-OE lines showed enhanced sensitivity (Kumar et al., 2017).

An additional layer of regulation comes from FERONIA (FER), a positive regulator of auxin-promoted growth and rapid alkalinisation factor (RALF) peptide signal. FER kinase is activated by phosphorylation enhanced by RALF (Chen et al., 2016). Then, FER interacts with guanine exchange factor 1 (GEF1), GEF4 and GEF10 that, in turn, activate GTPase ROP11/ARAC10, which physically interacts with ABI2, enhancing its activity and thereby linking the FER pathway with the inhibition of ABA signaling (Yu et al., 2012). Additionally, FER can interact with ABI2, which dephosphorylates FER and provides a feedback mechanism for RALF activation of FER (Chen et al., 2016). *A. thaliana* mutants disrupted in any step of the FER pathway displayed enhance sensitivity to ABA signaling.

The post-translational regulation can also affect the kinases. BRASSINOSTEROID INSENSITIVE 2 (BIN2) is an ARABIDOPSIS GLYCOGEN SYNTHASE KINASE 3 (GSK3)-like kinase, that can phosphorylate SnRK2.2 and SnRK2.3. Single *bin2* mutant plants are less sensitive to ABA in primary root inhibition, whereas a gain-of -function mutation of BIN2 shows enhanced sensitivity to ABA. This suggest a positive role of BIN2 phosphorylation of SnRKs in the ABA pathway (Cai et al., 2014). CLADE-E GROWTH-REGULATING 2 (EGR2) is a clade E phosphatase that interacts and inhibit OST1 in the plasma membrane. EGR2 is myristoylated by N-MYRISTOYLTRANSFERASE 1 (NMT1), and therefore, targeted to the membrane. Under cold stress, the interaction between EGR2 and NMT1 is attenuated, thus EGR2 cannot be myristoylated. This results in release of newly synthesized EGR2 from the plasma membrane, and therefore, OST1 can be activated in response to cold stress (Ding et al., 2019).

SnRKs can also be ubiquitinated, although this topic is starting to be studied. PHLOEM PROTEIN 2-B11 (AtPP2-B11) is a F-box protein that is part of the SKP1/Cullin/F-box E3 ubiquitin ligase complex. This ubiquitin ligase is able to interact and promote the degradation of SnRK2.3. The expression of AtPP2-B11 is also induced by ABA, suggesting that ABA can promote the degradation of SnRK2.3 to attenuate ABA signaling (Cheng et al., 2017). Further experiments should be conducted with α-SnRK2.3 antibodies to detect the degradation of the endogenous protein.

The ubiquitin system.

Ubiquitin is a 76-amino acid polypeptide conserved among animals, plants and fungi (Callis & Vierstra, 1989). Ubiquitin is encoded in different genes as homomeric or heteromeric fusions. Homomeric fusions are multimers of ubiquitin repeats head-to-tail. Heteromeric fusions encode ubiquitin followed by either small ribosomal proteins or by an ubiquitin-like protein called RUB (RELATED TO UB). These fusions are unable to function until the translation products are cleaved (Callis, 2014). In *A. thaliana* there are twelve ubiquitin encoding genes. The first of them, *UBQ1*, is also called *ERD16* (EARLY RESPONSE TO DROUGHT) because it was isolated as a mRNA induced in leaves after 1 hour of drought treatment (Kiyosue et al., 1994). The structure of different ubiquitins have been solved (**Figure 3**) (Hua & Vierstra, 2011). Ubiquitin first loop is able to adopt different conformations to interact with different ubiquitin binding proteins (Lange et al., 2008), while another region, called Ile-44 hydrophobic patch, interacts with the 26S proteasome as well as with other ubiquitin binding proteins (Sloper-Mould et al., 2001). The 26S proteasome is a proteolytic complex that breaks down ubiquitinated proteins but releases the attached

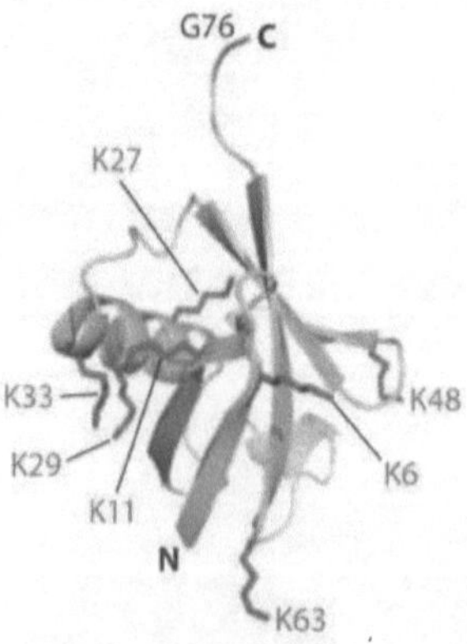

Figure 3. Three dimensional ribbon model of plant Ub. The side chains from the seven lysines in Ub that can be used for poly-Ub chain formation are shown in red. The β strands are in green, the α helices are in cyan, and the C-terminal Gly76 used to ligate Ub to other proteins is indicated. N, N terminus; C, C terminus. Reprinted from (Hua & Vierstra, 2011).

ubiquitins intact for reuse. The C-terminal Gly76 carboxyl group of the ubiquitin can form an isopeptide bound with a free lysyl ε-amino group of another protein or another ubiquitin (Hua & Vierstra, 2011). Therefore, proteins can be modified by the addition of a single ubiquitin (monoubiquitination), single ubiquitins at multiple sites of the same protein (multi-monoubiquitination) or ubiquitin chains (polyubiquitination) (**Figure 4**) (Akutsu et al., 2016). In an ubiquitin chain, ubiquitin moieties can be conjugated through one of their lysine residues (K6, K11, K27, K29, K33, K48 and K63) (**Figure 3**) or the N-terminal methionine residue (M1), offering countless possibilities to assemble a specific polymer. Ubiquitin chains that comprise only a single linkage type are called homotypic. In contrast, heterotypic chains contain mixed linkages within the same string. The assortment of ubiquitin chains used *in vivo* further increases given that heterotypic chains can also be branched – i.e. one ubiquitin molecule is ubiquitinated at two or more sites (Meyer & Rape, 2014). Additionally, ubiquitin chains adopt

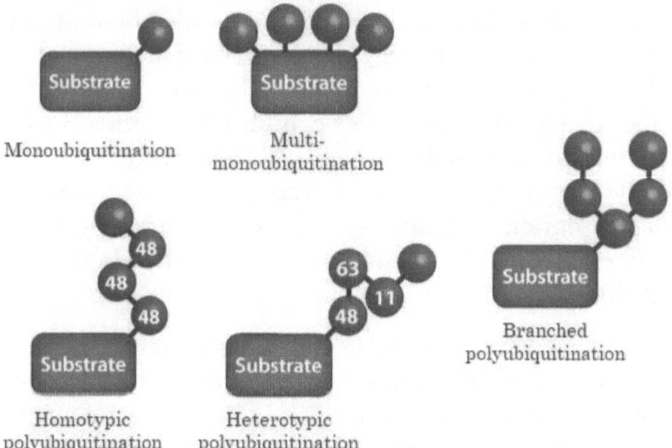

Figure 4. The different topologies of ubiquitination. Monoubiquitination, multi-monoubiquitination and polyubiquitination (homotypic and heterotypic). Branched polyubiquitination is a subtype of heterotypic polyubiquitination. Adapted from (Komander & Rape, 2012).

either "compact" conformations, where adjacent moieties interact with each other, or "open" conformations, where no interfaces are present except for the linkage site. The canonical K48-linked chains adopt compact conformations (Tenno et al., 2004). In the prevalent model for K48-linked diubiquitin, the ubiquitin moieties interact via their Ile44 patches, and two such diubiquitin modules pack tightly in tetraubiquitin (Eddins et al., 2007). However, this interaction can also happen between the Ile36 patch of one ubiquitin with the Ile44 patch of the proximal unit. This structural flexibility might give binding partners of K48-linked chains access to the Ile44 patch, a hot spot for ubiquitin recognition (Komander & Rape, 2012). Similar to K48 linkages, K6- and K11-linked chains adopt compact conformations, with K11-linked chains also displaying structural flexibility to interact with the Ile44 patch (Bremm & Komander, 2011). In contrast, Met1- and K63-linked chains mostly display open conformations. Therefore, most binding partners exploit the distance and flexibility between chain moieties, rather than recognizing a defined geometric assembly of different ubiquitin surfaces (Sims & Cohen, 2009). Together, the various ubiquitination patterns reveal a large array of geometries that can be utilized by binding partners to distinguish between modifications. Moreover, ubiquitin molecules can also be modified by other post-translational modifications, including acetylation, which inhibit chain elongation (Ohtake et al., 2015), and phosphorylation (Swaney et al., 2015), representing yet another layer of ubiquitin signal regulation and/or diversification.

The ubiquitin system is the intracellular protein modification pathway that attaches one or more ubiquitins to a target protein localized in different subcelular compartments. The attachment of ubiquitin leads to proteolytic and non-proteolytic fates. This modification can alter not only protein half-life, but also its activity, interaction or localization (Callis, 2014; Dupre & Haguenauer-Tsapis, 2001; Hua & Vierstra, 2011). The process of ubiquitination requires three distinct biochemical activities catalysed by enzymes denominated E1, E2 and E3 ubiquitin ligases (**Figure 5**) and the hydrolysis of ATP (Ciechanover et al., 1980). The first enzyme (E1) activates the ubiquitin by adenylation of its C-terminal carboxyl group, and then, the ubiquitin is attached to a cysteinyl residue of the E1 by a thioester covalent linkage. E1 activity is encoded in *A. thaliana* by one of two related genes *UBIQUITIN ACTIVATING 1* (*UBA1*) and *UBA2*.

In the second step the activated ubiquitin is transferred to a cysteinyl residue of the E2, forming a new thioester covalent linkage. E2s can vary in overall length, but their ubiquitin conjugating domain (UBC) has a conserved HPN (His-Pro-Asn) motif that plays a catalytic role by stabilizing the transient oxyanion formed during ubiquitination transfer (Wu et al., 2003). The *A. thaliana* genome encodes 48 UBC domain containing proteins; however, only 37 carry the active-site cysteine required for E2 Ub-conjugase activity (Callis, 2014).

Some E2s are specialized in substrates targeted by specific E3s, like UBC19 and UBC20, that constitute the subfamily VIII and work together with the E3 APC/C (Criqui et al., 2002). However, most E2s display high and promiscuous activity with different E3s (Bueso et al., 2014b; Kowarschik et al., 2018; Kraft et al., 2005).

The next step is the transfer of the ubiquitin to the substrate, which is recognize by E3 ligases. The ubiquitin can be transferred directly from the E2 or by forming an intermediate thioester covalent linkage with the E3 (Callis, 2014), before forming the isopeptide bound with a free lysyl ε-amino group of the substrate or another ubiquitin already linkage to the substrate to form a ubiquitin chain. Moreover, E2s also play a key role to mediate ubiquitin chain assembly on the target (Callis, 2014; Kowarschik et al., 2018). In the case of U-box/RING-type E3 ligases, concerted action of the E2/E3 pair regulates substrate specificity and the type of Ub modification on the substrate proteins (Zhao et al., 2013). In the case of mammalian HECT- and RBR-type E3s, residues in

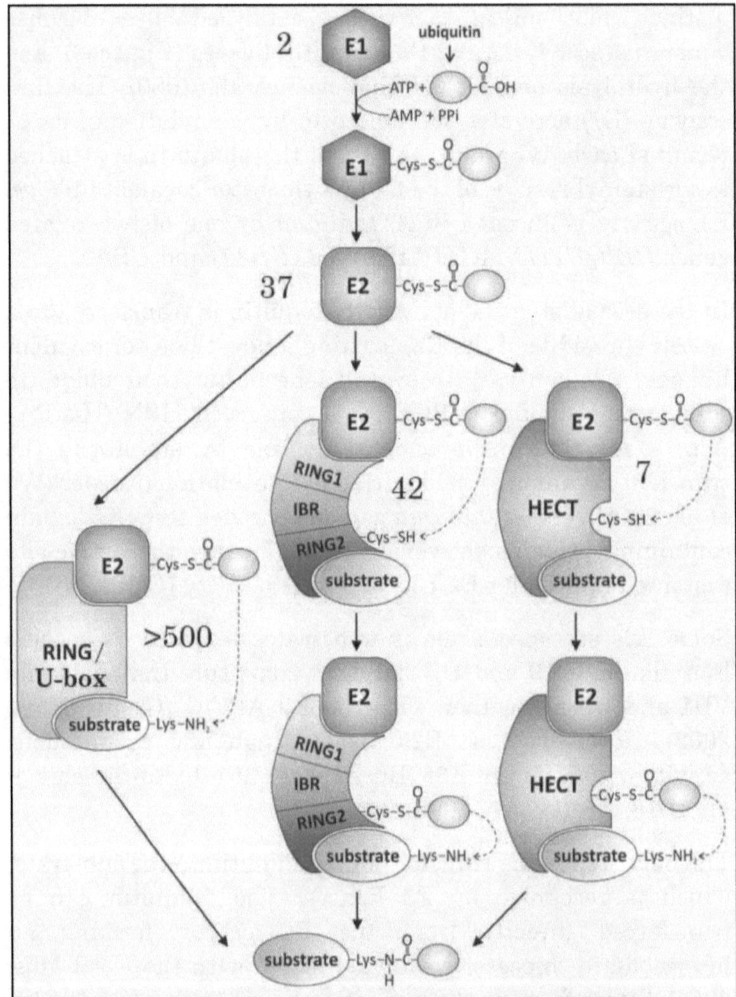

Figure 5. Ubiquitin genes and ubiquitination pathway. Ubiquitin is activated by E1, and thioester conjugated first to E1, then to E2. E2~Ub interacts with an E3. In the case of RING and U-box E3s, an intermediate complex of substrate, E3 and E2~Ub is required for transfer to the substrate. For RBR and HECT-type E3s, E2~Ub interacts and transfers ubiquitin to an E3 cysteinyl sulfhydryl prior to its transfer to the substrate. Only monomeric E3s are represented. Numbers represent number of genes of each element of the pathway in *A. thaliana*. Adapted from (Winklhofer, 2014).

the active site Cys determine the type of ubiquitination, although little is known for plant enzymes (Dove et al., 2016; Kim & Huibregtse, 2009).

The most common chains in *A. thaliana* are formed from the K48, while K63 and K11 are also quite common. Much scarce are the chains linked via other lysines. K27 is the only non-surface exposed lysine, thus, linkage assembly would require localized changes in ubiquitin structure (Kim & Kim, 2013). The ubiquitination pattern dictates downstream events, like polyubiquitination, and specially K48-linked polyubiquitination, which promotes 26S proteasome degradation, whereas monoubiquitination promotes internalization of plasma membrane proteins and K63-linked polyubiquitination can promote subsequent endosome trafficking (Romero-Barrios & Vert, 2018; Yu & Xie, 2017). A single ubiquitin is sufficient to trigger recognition by the clathrin-interacting Epsin/Eps15-like endocytic adaptors and internalization (Haglund et al., 2003; Shih et al., 2002; Shih et al., 2000). However, the presence of K63-linked Ub chains increased avidity for Ub-binding adaptors boosting endocytosis rates (Galan & Haguenauer-Tsapis, 1997; Paiva et al., 2009; Sato et al., 2009b; Sims et al., 2009). Internalized ubiquitinated proteins are subjected to recycling on deubiquitination, or proceed towards the degradative pathway to the vacuole/lysosome. Ubiquitination/de-ubiquitination is a key process for intracellular trafficking. (Tanno & Komada, 2013; Tian & Xie, 2013). Therefore, ubiquitination is a way to change protein location (Callis, 2014). De-ubiquitination is performed by a big family of proteins called deubiquitinating enzymes (DUBs) (March & Farrona, 2017). DUBs cleave the conjugated ubiquitin from its substrates, modulating the stability, activity or destiny of their target proteins. DUBs are subdivided in two major types: cysteine proteases named UBIQUITIN-SPECIFIC PROTEASES (UBPs) and metalloproteases (Zhou et al., 2017).

Before reaching the vacuole/lysosome, K63 polyubiquitinated proteins are sorted into intraluminal vesicles (ILVs) by the ENDOSOMAL SORTING COMPLEXES REQUIRED FOR TRANSPORT (ESCRT) complex (Hurley & Ren, 2009). Five ESCRT complexes have been described in eukaryotes; the ESCRT-0, -I, -II, -III, and ESCRT-III-associated AAA ATPase SKD1/Vps4 complexes, although plants seem to lack canonical ESCRT-0 subunits (Paez Valencia et al., 2016; Richardson et al., 2011; Winter & Hauser, 2006). Interestingly, plant-specific ESCRT components have been characterized (Belda-Palazon et al., 2016; Gao et al., 2014; Gao et al., 2015; Kolb et al., 2015; Reyes et al., 2014). For example, ESCRT-I complexes, FYVE1 and VPS23A proteins are known to interact with each other and are required for recognition and internalization of PYR/PYL receptors into ILVs (Belda-Palazon et al., 2016; Shen et al., 2016). Additionally, activity of the ESCRT-III-associated protein ALG-2 INTERACTING PROTEIN-X (ALIX) is required for PYR/PYL internalization (Garcia-Leon et al., 2019). ALIX physically interacts with FYVE1 and VPS23A acting as a bridge between the ESCRT-I and -III complexes and thereby coordinates ubiquitinated proteins sorting and vacuolar targeting. During this sorting process towards vacuole/lysosomes, K63-linked chains are removed to recycle Ub moieties. Therefore, DUBs are also required for proper sorting of the ubiquitinated proteins (Dupre & Haguenauer-Tsapis, 2001; McCullough et al., 2004; Mizuno et al., 2006; Row et al., 2006). Finally, ILVs will be released into the vacuolar/lysosomal lumen upon fusion of the multivesicular bodies with the tonoplast, and associated proteins will be degraded (Romero-Barrios & Vert, 2018).

An ubiquitin-dependent proteolytic pathway recognizes the N-terminal residue of proteins and was initially characterized in yeast and mammalian cells (Tasaki et al., 2012; Varshavsky, 2011). It is name the N-degron pathway, and studies have demonstrated that the nature of the N-terminal residues of a protein determines its stability. Subsequent studies have revealed its presence in plants (Graciet et al.,

2010; Zhang et al., 2018a). For example, in A. thaliana, the family of transcription factors ERFVII is involved in physiological responses to hypoxic environmental conditions and conserved a N-terminal degron motif (Dissmeyer, 2019), which plays a central role in protein turnover. Additional to the N-degron pathways, there are also C-degron pathways that are related functionally. Instead of being in the N-terminal residues, C-degrons are degradation signals whose main determinants are the C-terminal residues (Varshavsky, 2019). This field is much more recent, but in 2.018 a large set of C-degrons was found in human proteins (Chatr-Aryamontri et al., 2018; Koren et al., 2018; Lin et al., 2018).

E3 ubiquitin ligases.

The third activity in the ubiquitin conjugating cascade is known as E3 or ubiquitin ligase and facilitates the transfer of ubiquitin to the substrate protein. E3s are a large group, more than 1500 have been predicted in *A. thaliana* (Hua & Vierstra, 2011), which can be divided in monomeric or multimeric E3 ubiquitin ligases based on their number of polypeptide chains.

Monomeirc E3 ligases.

Monomeric E3s are subdivided into two types based on whether the ubiquitin is passed directly from the E2 to the substrate or they carry the ubiquitin through a thioester covalent linkage on an active cysteine (Callis, 2014). The first subtype includes the U-box- and RING-type E3s, and the second, the HECT- and RBR-type E3s (**Figure 5**).

1. RING-type E3 ubiquitin ligases.

The RING-type E3s are the most numerous in *A. thaliana*, there are 490 proteins with a RING domain proposed to function as E3s. The RING-type of E3 ligases shares the RING domain, a ~40–60 amino acid region containing an octet of spatially conserved cysteine and histidine residues that bind two Zn atoms in a unique "cross-brace" arrangement (**Figure 6**). The spacing of the Cys/His residues is different to other Zn-binding "fingers", such as those of Zn finger transcription factors, resulting in a structural difference between them. This domain serves as the major E2-interacting region (Callis,

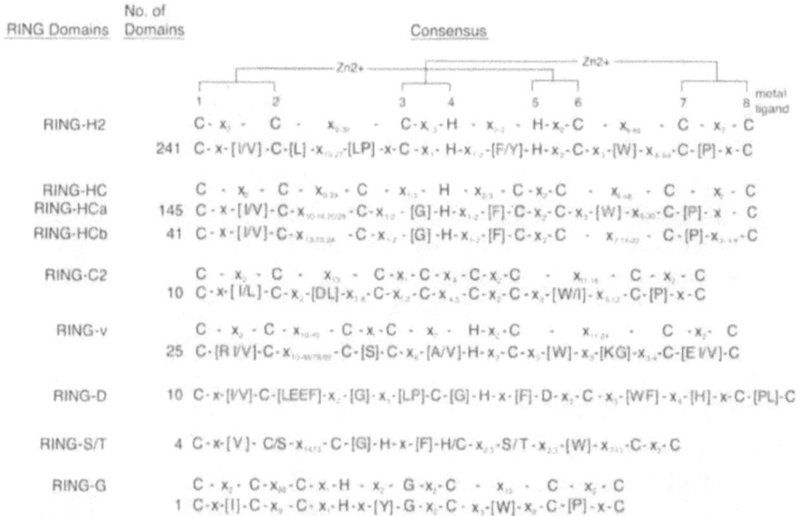

Figure 6. Consensus and number of each type of RING domain identified in A. thaliana. Consensus previously characterized canonical RING domains (RING-H2 and RING-HC) and modified RING domains (RING-C2, RING-v, and RING-G) are shown above that of Arabidopsis. The conserved metal ligand positions and zinc (Zn21) coordinating amino acid pairs are illustrated using the RING-H2 consensus. The eight conserved Cys (C) and His (H) residues of each RING type and metal ligand substitutions in the RING-D, RING-S/T, and RING-G domains are shown. Other conserved amino acids are shown in brackets. For each RING type, amino acids in red are found in ≥ 80% of domains, and blue and green in ≥ 50% and ≤ 50% of domains, respectively. $X_{(n)}$ indicates number of spacing amino acids between conserved amino acids. Reprinted from (Stone et al., 2005)

2014). Given the high number of RING-type E3 ligases, they have been classified according to protein-protein interacting motifs and the distance between each of the eight zinc-coordinating Cys and/or His residues, this way RING-type E3 ligases have been divided into 30 different groups (Stone et al., 2005). The biological functions of RING-type E3 ligases are very different, and there are more functions to be elucidated. As previously described, RGLG1 and RGLG5 are RING-type E3 ligases that target PP2CA for degradation, and since PP2CA is a negative regulator of ABA signaling, RGLG1 and RGLG5 act as positive regulators of the pathway (See below, **RING DOMAIN LIGASE (RGLG): RING-type E3 ubiquitin ligase family**).

RING-type E3 ligases also regulate ethylene biosynthesis, since ACC SYNTHASE 4 (ACS4) and ACS7 catalyse the rate-limiting step in ethylene biosynthesis and are both ubiquitinated by XBAT32 (Prasad & Stone, 2010). Other representatives of the family are the 91 ATLs (Arabidopsis genes toxic to yeast), that have a RING domain and also a transmembrane domain (Kim & Kim, 2013). Many ATLs are implicated in biotic and abiotic stress responses. For example, *ATL2* and *ATL9* mRNAs accumulate in response to chitin (Berrocal-Lobo et al., 2010), while *ATL78* mRNA is increased in response to cold (Kim & Kim, 2013). Moreover, SUPPRESSOR OF PLASTID PROTEIN IMPORT LOCUS (SP1) regulates the abundance of TOC (Translocation at the outer envelope of chloroplasts) complexes, the outer membrane translocation machinery of chloroplasts (Ling et al., 2012). SP1 is anchored to the outer envelope via two TM regions, with its C-terminal RING domain facing the cytosol. SP1 interacts and is able to ubiquitinate multiple TOC proteins *in vitro*. This turnover of TOC is proposed to facilitate developmental transitions such as greening and senescence.

RING proteins can function as single polypeptides, however, they can also interact with each other or related RINGs to form homomeric or heteromeric complexes (Callis, 2014). In *A. thaliana*, HUB1 and HUB2 are suggested to function together as a heteromeric ligase for mono-ubiquitination of histone H2b (Feng & Shen, 2014). Furthermore, interaction with other proteins has also been described, like adaptor proteins that function to recognize substrates, modulate their activity or to localize the ligase to a particular subcellular compartment or attach it to a membrane surface (Callis, 2014).

2. U-box-type E3 ubiquitin ligases.

U-box and RING E3 ligases show a very similar mechanism of action, but differ in the residues that conform the E2 docking site. However, the structure of the U-box and RING domains show a similar fold. In the RING motif there are cysteine and histidine residues that chelate Zn^{2+}, but in the U-box are replaced by a network of hydrogen bonds using cysteine, serine and glutamate side chains. The tertiary structure is also stabilized by hydrophobic interactions and salt bridges (Andersen et al., 2004). There are around 60 U-box proteins identified in *A. thaliana* (Callis, 2014), which have been defined PUB proteins and consecutively numbered after the term PUB, with the exception of CARBOXYL TERMINUS OF HSC70-INTERACTING PROTEIN (CHIP) (Ryu et al., 2019).

PUBs are involved in self-incompatibility, hormone responses, defence and abiotic stress responses (Yee & Goring, 2009). PUB22/23 were reported to degrade PYL9, while PUB12/13 target ABI1. Additionally, PUB12/13 catalyse ubiquitination of FLAGELLIN SENSING2 (FLS2) and *pub12 pub13* loss-of-function plants are more resistant to *Pseudomonas syringae* than wild type plants (Lu et al., 2011). Interestingly, mRNAs for several PUBs increased in response to ABA, PUB19 increased ~160-fold after 3–5 hours in 50 µM ABA (Hoth et al., 2002). *pub19* and *pub18* loss-of-function plants showed enhanced ABA sensitivity, while PUB19 and PUB18 OE plants showed reduced ABA sensitivity (Liu et al., 2011; Seo et al., 2012).

3. HECT-type E3 ubiquitin ligases.

There are 7 HECT E3s in *A. thaliana* named UBIQUITIN PROTEIN LIGASES (UPLs). UPLs contain a C-terminal HECT domain that accepts ubiquitin from an E2 conjugating enzyme through a thioester covalent linkage on an active cysteine and then transfers the ubiquitin to the target substrate (Furniss et al., 2018). N-terminal to the HECT domain, UPLs contain different interaction motifs.

UPL1, UPL3, UPL4 and UPL5 function as regulators of SA-responsive genes expression and immunity (Furniss et al., 2018). UPL3 has been reported to regulate trichome branching by targeting transcription factors GLABROUS 3 (GL3) and ENHANCER OF GL3 (EGL3), which control trichome development and flavonoid metabolism (Downes et al., 2003; Patra et al., 2013). UPL5 targets WRKY53, a transcription factor that regulates leaf senescence (Zentgraf et al., 2010). Moreover, in *Brassica rapa* there are 10 HECT genes described, which all were responsive to salt and drought stresses (Alam et al., 2019).

4. RBR-type E3 ubiquitin ligases.

There are 42 RBR E3s in *A. thaliana* (Marin, 2010), which are structurally characterized by the presence of three putative RING domains in tandem, named as RING1, IN BETWEEN RING (IBR) and RING2 (Callis, 2014; Marin, 2010). Studies done in mammals have showed that RBR E3s combine properties of both RING- and HECT-type E3s (Wenzel & Klevit, 2012; Wenzel et al., 2011). Non covalent interaction with E2 occurs at the RING1 domain as in RING- and U-box E3s, but then the activated ubiquitin is transferred to a conserved cysteine residue in RING2. Finally, this cysteinyl residue of RING2 transfers the ubiquitin to a substrate in a HECT-type mechanism (Callis, 2014). RING1 is the only domain with a cross-brace zinc-coordination topology, whereas the zinc-liganding residues in IBR and RING2 domains are arranged in a sequential fashion (Duda et al., 2013; Riley et al., 2013). Therefore, the RING nomenclature for IBR and RING2 does not reflect actually the canonical

RING cross-brace structure, which is only present in RING1. Molecular insights into the function of RBRs have been obtained mostly in humans from the study of Parkin, associated with autosomal recessive Parkinsonism, and Human Homologue of Ariadne (HHARI), involved in regulation of translation (Dove et al., 2016).

One prominent subgroup is the Ariadne (ARI) E3s, which are 14 members and 2 pseudogenes. ARI12 and ARI14 have been implicated in UV-B signaling and fertilization, respectively (Lang-Mladek et al., 2012; Ron et al., 2010), but the rest of the family is hardly characterized. *GmARI1* from soybean in *A. thaliana* enhances aluminium tolerance (Zhang et al., 2014). Additionally, RSL1 belongs to a gene family composed of at least 9 more members: RING finger ABA-related 1 (RFA1) to RFA9. Initially, RSL1 was annotated as a RING-type E3 ligase (Bueso et al., 2014b); however, further inspection of the gene family revealed that RSL1/RFAs belong to the RBR-type E3 ligase family.

RING DOMAIN LIGASE (RGLG): RING-type E3 ubiquitin ligase family.

This family is composed of five members, i.e. RGLG1 to 5 (Yin et al., 2007). Sequence alignment has revealed that RGLG1, RGLG2 and RGLG5 belong to a different branch than RGLG3/4, which are involved in jasmonate-mediated responses (Zhang et al., 2012). RGLG1 and RGLG2 have been reported to affect hormone signaling since the *rglg1 rglg2* loss-of-function plants showed altered auxin and cytokinin levels (Yin et al., 2007). Moreover, RGLG1 and RGL2 show functional overlapping in K63-linked polyubiquitination in plasma membrane, which is a proteasome-independent function (Romero-Barrios & Vert, 2018). Indeed, the *rglg1 rglg2* double mutant shows reduced levels of Ub-K63 chain-specific signals for PIN2 auxin carrier, which affects its turnover and the root-hair phenotype of iron deficient plants (Leitner et al., 2012; Li & Schmidt, 2010).

For ABA signaling, it has been demonstrated that ABA induces the degradation of PP2CA through the RGLG1/5 E3 ligases and the Ub 26S proteasome system. ABA receptors were not required for the *in vitro* ubiquitination of PP2CA by RGLG1/5; however, under *in vivo* conditions ABA enhanced the interaction of RGLG1/5 with PP2CA through an unknown mechanism (Wu et al., 2016). In contrast, ABI1, which acts as an ABA co-receptor in the ternary ABA-ABI1-PYR/PYL complex (Miyazono et al., 2009; Yin et al., 2009), is degraded through an ABA- and PYR1-dependent mechanism that triggers recognition by PUB12/13 E3 ligases (Kong et al., 2015).

However, RGLG2 does not interact with clade-A PP2Cs (Wu et al., 2016), so is functionally unrelated to RGLG1/5 regarding the 26S proteasome-linked polyubiquitination of PP2Cs. RGLGs may have functional diversity due to their variable N-terminal regions. Moreover, RGLG1 and RGLG2 have different expression profiles in response to different stimuli. RGLG1 is highly induced by ABA, in contrast to RGLG2 (**Figure 7**). Additionally, RGLG1 is induced by cold and osmotic and salt stresses. In contrast, RGLG2 does not respond to cold and only responds to osmotic and salt stress in the roots (**Figure 8**). RGLG1 OE enhances drought tolerance due to enhancing PP2CA degradation (Wu et al., 2016). In contrast, RGLG2 negatively regulates the drought stress response (Cheng et al., 2012). Both RGLG1 and RGLG2 are plasma membrane-associated proteins with a predicted N-terminal myristoylation site (Yin et al., 2007) (**Figure 9**).

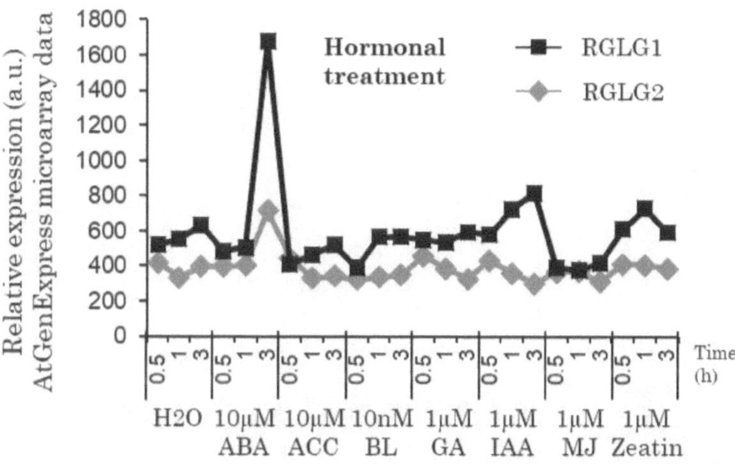

Figure 7. Relative expression of RGLG1 and RGLG2 in seedlings. Seedlings of A. thaliana were treated without (H2O) or with different hormones: 10mM ABA, 10mM 1-aminocyclopropane-1-carboxylic acid (ACC), 10mM brassinolide (BL), 10mM gibberellin (GA), 10mM indole-3-acetic acid (IAA), 10mM methyl jasmonate (MJ) or 10mM zeatin. Expression data was obtained from Affymetrix microarrays for A. *thaliana* deposited into the AtGenExpress public database (Goda et al., 2008), visualized using the AtGenExpress Visualization Tool located in http://weigelworld.org/resources.html.

In presence of stress RGLG2 translocate to the nucleus, where contributes to the degradation of AtERF53, a transcription factor that positively regulates drought-responsive genes. This can possibly be a desensitization mechanism. Different enzymes and signaling proteins shuttle to the nucleus in response to environmental cues or stress, which enables interactions with their targets (Bigeard & Hirt, 2018; Lee et al., 2015). RGLG1 is also localized in the plasma membrane, where mediates K63-linked polyubiquitination and the endocytic turnover of plasma-membrane transporters. However, PP2CA is predominantly localized in the nucleus and degraded via the 26S proteasome (Wu et al., 2016). For this reason, we hypothesized that ABA treatment might modify the subcellular localization of RGLG1 and promote its interaction with PP2CA.

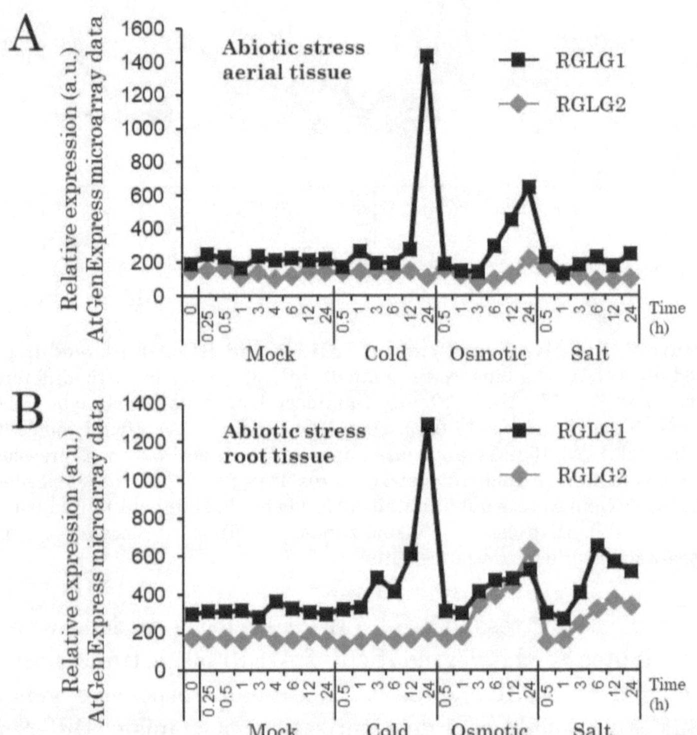

Figure 8. Relative expression of RGLG1 and RGLG2 in seedlings.
Seedlings of A. thaliana were submitted to cold (4°C), osmotic (300mM mannitol) or salt (150mM NaCl) stress. Expression data was obtained from Affymetrix microarrays for *A. thaliana* aerial tissue **(A)** or root tissue **(B)** deposited into the AtGenExpress public database (Goda et al., 2008), visualized using the AtGenExpress Visualization Tool located in http://weigelworld.org/resources.html.

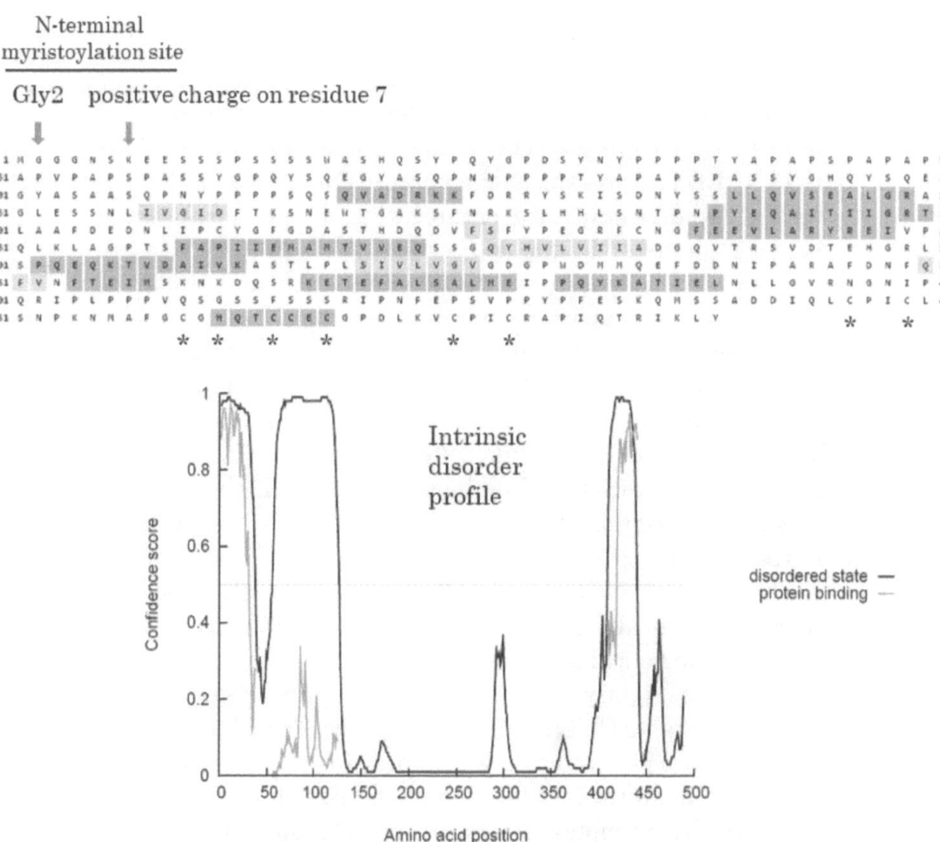

Figure 9. Scheme of the RGLG1 protein. The position of the N-terminal myristoylation site, secondary structure prediction (pink, a-helix; yellow, b-sheet), the intrinsic disorder profile (bottom) and the RING finger (conserved Cys and His residues labelled by asterisks) are indicated. Analysis was performed using PSIPRED and DISOPRED3 prediction servers (http://bioinf.cs.ucl.ac.uk). The graph (bottom) shows the DISOPRED3 disorder confidence levels against the sequence positions as a solid blue line. The grey dashed horizontal line marks the threshold above which amino acids are regarded as disordered. For disordered residues, the orange line shows the confidence of disordered residues being involved in protein-protein interactions, i.e. the disordered region folds upon binding to an interacting protein.

Some proteins with a RING domain are components of multimeric complexes. The most notable example is RING BOX 1 (RBX1), present in the cullin-RING based E3 ligases (CRLs), whose structural organization is highly conserved between plants and animals (Lechner et al., 2002). CRLs are multimeric E3 ligases, where E2 interaction and substrate binding occur in different subunits tethered together into a single complex by an elongated cullin (CUL) scaffold protein (Hua & Vierstra, 2011). Studies using mammalian cells suggest that at least 20% of all proteasome-mediated degradation is CRL dependent (Soucy et al., 2009). In plants there are 3 main CUL families: CUL1, CUL3 and CUL4 (**Figure 10**, top). In *A. thaliana* there are 6 CUL subdivided in the previous groups: CUL1/CUL2a/b, CUL3a/b and CUL4. Each different CUL assembles a distinct CRL collection (Gingerich et al., 2005). The C-terminal region of CUL contains two helical domains that create a V-shaped cleft; this cleft becomes tightly occupied by the N-terminal helix of RBX1 to create a stable CUL/RBX1 catalytic core. The RING domain in RBX1 provides a docking platform for E2s charged with activated ubiquitin and binding of the E2-Ub to this pocket allosterically promotes transfer of the ubiquitin from the E2 directly to the substrate (Kleiger et al., 2009). RBX1 is considered an essential gene; homozygous loss-of-function seedlings are not viable (Lechner et al., 2002). Down-regulation of RBX1 generates dwarf plants with poor fertility, indicating the central role that CRLs play in plant growth and development.

At the opposite site of the RBX1 binding site, the N-terminal bundle of CUL terminates in a hydrophobic/polar patch that binds specific sets of substrate adaptors, which create a large pocket to grasp the substrate. Five major groups of CRLs have been described in metazoans based on their substrate adaptor, but only three have been described in plants, which are known

CRLs

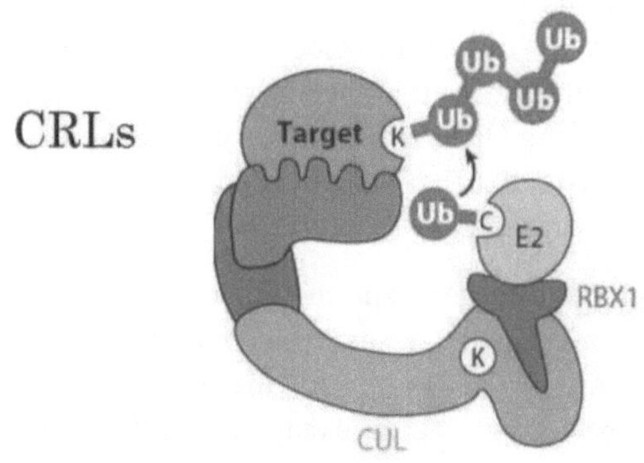

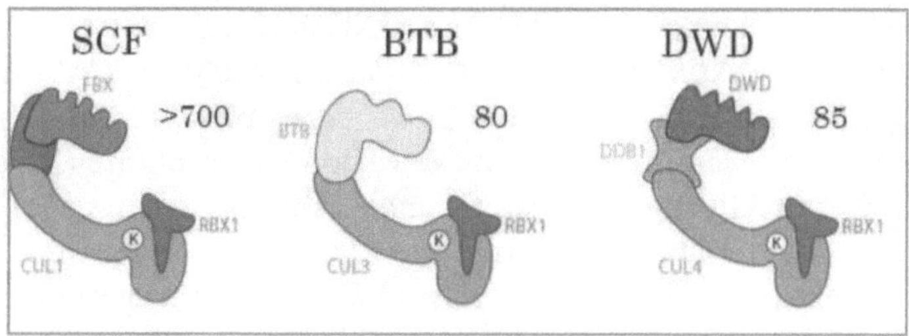

APC/C

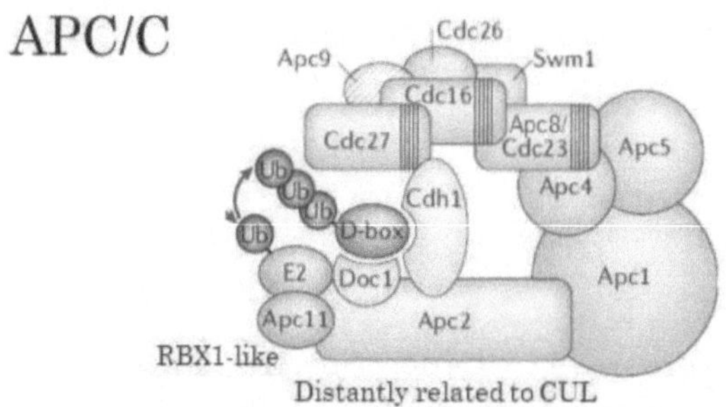

Figure 10. Structure of plants multimeric E3 ligases. The three types of CRL (CRL1^{SCF}, CRL3^{BTB}, CRL4^{DWD}) (top) and the APC/C (bottom). Numbers represent number of genes of each element of the pathway in *A. thaliana*. Adapted from (Hua & Vierstra, 2011) (top) & (Peters, 2006) (bottom).

as SKP1-Cullin-1-F-box protein (SCF), CRL3^{BTB} and CRL4^{DWD} complexes (Hua & Vierstra, 2011). Upon interaction with the substrate certain lysine residues are oriented to the activated ubiquitin. Thus, simple positioning of one or more accessible substrate lysines in a "hot zone" close to the E2 may be sufficient to drive ubiquitin transfer (Cardozo & Pagano, 2004). Once bound to the CRL, the substrate is ubiquitinated with at least four ubiquitin moieties needed to direct the substrate to the 26S proteasome (Pickart & Fushman, 2004).

Besides CRLs there is another type of multimeric E3 ligase, the Anaphase promoting complex/cyclosome (APC/C). In plants, the APC/C consists of at least 11 core subunits, each of which is encoded by a single gene, except for APC3 (**Figure 10**, bottom). APC2, a CUL-like protein, functions as the scaffold of the complex, while the RING domain is provided by APC11, a relative of RBX1 (Fulop et al., 2005). APC/C core complex and its activators have been reported to play important roles in growth and development in plants (Xu et al., 2019). APC8 is required for male meiosis, but in general there is not much information about this type of multimeric E3 ligase in plants. Interestingly, *O. sativa* ABA receptors are targeted by the APC/C complex (Lin et al., 2015) as we previously mentioned (See above, **Regulation of ABA core components by post-transcriptional mechanisms and interacting proteins**).

1. Cullin 1-RING based E3 ligase (CRL1).

One of the three groups of CRLs described in plants is the SCF complex, which consist of the CUL1 isoform, RBX1 and a substrate adaptor. The first substrate adaptor discovered (the one that gave the group its name) was the S-phase kinase-associated protein (SKP)-1/cyclin-F heterodimer (Feldman et al., 1997; Skowyra et al., 1997). Cyclin-F is part of a large protein family distinguished by a common N-terminal domain called the F-box (FBX) (Bai et al., 1996). FBX sequence assumes a compact trihelical fold that forms an interlocked interface with a broad shallow pocket in SKP1, and them the heterodimer SKP1/FBX docks into the N terminus of CUL1 (Zheng et al., 2002). In *A. thaliana* there are ~700 functional *FBX* genes and ~200 pseudogenes, which combined with 19 *SKP1* genes (*ASK1-19*) could assemble thousands of different CRL^{SCF} complexes (Gagne et al., 2004).

Some FBX proteins work as hormone receptors or coreceptors. For example, auxin signal is regulated by a family of auxin response factor (ARFs), which either promote or inhibit the transcription of a variety of auxin effector genes. In absence of auxin, ARFs are repressed by direct association with a large family of auxin/indole-3-acetic acid (Aux/IAA) repressors (Gray et al., 2001), together with TOPLESS corepressor (Hua & Vierstra, 2011). However, in presence of auxin, auxin is sensed by a co-receptor system that involves substrate adaptors subunits of CRL1, TRANSPORT INHIBITOR RESPONSE (TIR)-1 and AUXIN-BINDING FBX (AFB)-1-5 and the Aux/IAA repressors (**Figure 11**), thus Aux/IAA are ubiquitinated and degraded through the proteasome to release the auxin signalling (Dharmasiri et al., 2005; Gray et al., 1999; Gray et al., 2001; Peer, 2013). Auxin favours the formation of the co-receptor system interacting between the F-box protein and the Aux/IAA repressor as a 'molecular glue' (Del Pozo & Manzano, 2014; Tan et al., 2007).

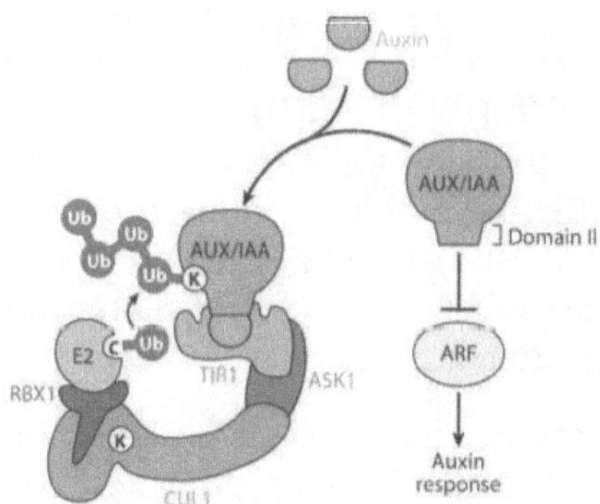

Figure 11. Structure of the SCF^{TIR1} auxin receptor. Diagram shows the organization and mechanism of action for the SCF^{TIR1} complex following auxin binding. Reprinted from (Hua & Vierstra, 2011).

Similar is the regulation of the JA pathway, JA-ZIM (JAZ) are transcriptional regulators that repress JA response by binding and inhibiting MYC2 transcription factor (Chini et al., 2007). CORONATINE-INSENSITIVE (COI)-1 is an FBX protein that works as the JA receptor, which forms a complex with jasmonoyl isoleucine (Ja-Ile) and JAZ proteins (**Figure 12**). As a result, JAZs are targeted for ubiquitination and degradation via the 26S proteasome, which leads to MYC2 activation (Sheard et al., 2010).

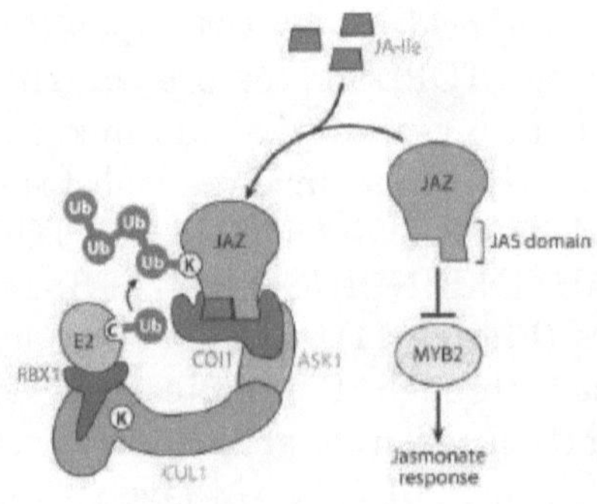

Figure 12. Structure of the SCF^{COI1} jasmonate receptor. Diagram shows the organization and mechanism of action for the SCF^{COI1} complex following Ja-Ile binding. Reprinted from (Hua & Vierstra, 2011).

The GA pathway also follows a relief of repression mechanism, but FBX proteins do not bind the hormone. GA signal is repressed by DELLA transcriptional repressors, which restrain plant growth. GA is recognized by GIBBERELLIN INSENSITIVE DWARF1 (GID1), which then interacts with the DELLA proteins (Salazar-Cerezo et al., 2018). The complex GID1-GA-DELLA is recognized by SCF substrate adaptors SLEEPY (SLY)-1 and SNEEZY (SNE)-1, and DELLA proteins are targeted for degradation via the 26S proteasome (Shimada et al., 2008).

Finally, in the ethylene pathway happens the opposite, the SCF complexes maintain the signal off by promoting the degradation of ETHYLENE-INSENSITIVE (EIN)-3 and EIN3-LIKE (EIL)-1 transcription factors, and the endoplasmic reticulum EIN2 signaling factor. EIN3 and EIL1 are recognized by EIN3-binding FBX protein (EBF)-1/2 (Konishi & Yanagisawa, 2008), while EIN2 is recognized by EIN2-targeting protein (ETP)-1/2 (Qiao et al., 2009). EIN2 is also phosphorylated by CONSTITUTIVE TRIPLE RESPONSE (CTR)-1 Ser/Thr kinase in absence of ethylene, which keeps EIN2 inactive and promotes its degradation by ETP1/2. The presence of ethylene is detected by ETHYLENE RESPONSE (ETR)-1, which stops the phosphorylation of EIN2. EIN2 is then cleaved and its C-terminus fragment blocks the translation of EBF1/2, which stops EIN3 and EIL1 degradation (Merchante et al., 2013).

Besides hormone perception, some FBX proteins also work as photoreceptors. FBX protein ZEITLUPE (ZTL) is a SCF substrate adaptor that targets TIMING OF CAB EXPRESSION 1 (TOC1), pseudoresponse regulator (PRR)-5, and CYCLIN DOF FACTOR (CDF)-1 for ubiquitination in the dark part of the daily cycle. ZTL is a blue-light photoreceptor that has an N-terminal LOV (Light, Oxygen, or Voltage) domain. Perception of blue light alters its binding affinity for GIGANTEA (GI), which interacts with its LOV domain only under blue light stabilizing ZTL, and thus shows a protein peak at dusk. ZTL family members, FLAVIN BINDING KELCH REPEAT F-BOX1 (FKF1) and LOV KELCH PROTEIN 2 (LKP2), also contribute to control the pace and robustness of the circadian clock through the regulation of TOC1 and PRR5 protein stability (Thines et al., 2019). An additional function of SCF complexes is promoting incompatibility barriers that avoid self-pollination (Hua & Kao, 2006).

2. Cullin 3-RING based E3 ligase (CRL3).

Another group is the CRL3[BTB] complex, which is formed by a CUL3 isoform, RBX1 and a substrate adaptor that belongs to the Broad-complex, Tramtrack, and Bric-à-brac / POx virus and Zinc finger (BTB/POZ, hereinafter called BTB) superfamily. The BTB proteins contain a BTB fold to bind CUL3 and, additionally, a variety of protein-protein interaction domains to recognize the substrate (Zhuang et al., 2009). In plants, these domains are armadillo, ankyrin repeats, tetratricopeptide, nonphototropic-hypocotyl 3 (NPH3), meprin-and-TRAF-homology (MATH) and a variety of other sequence motifs (Gingerich et al., 2007). According to those domains, the 80 BTB proteins in *A. thaliana* have been classified into different clades (**Figure 13**).

The ethylene signaling pathway is regulated by come SCF complexes as we described above, and the ethylene biosynthesis is regulated by at least three BTB proteins, i.e. ETHYLENE OVERPRODUCER (ETO)-1, ETO1-LIKE (EOL)-1 and EOL2. These proteins are substrate adaptors of CRL3 complexes that recognize ACSs, which are the rate limiting enzymes of ethylene biosynthesis as we previously described (See above, **1. RING-type E3 ubiquitin ligases**) (Wang et al., 2004). In absence of ethylene, ETO1 and EOL1/2 recognize ACSs, but in presence of the hormone, ACSs are phosphorylated, which blocks the E3 recognition, creating a positive feedback that induces more ethylene production.

Other CRL3 E3 ligases act in downstream steps of photomorphogenesis. Three BTB proteins, NPH3 and LIGHT-RESPONSE BTB 1 (LRB1) and LRB2, play important roles in regulating blue- and red-light perception by ubiquitination of phototropin-1 (Pedmale & Liscum, 2007) and phytochrome-B photoreceptors (Hua & Vierstra, 2011), respectively. NPH3 protein is constituted by a BTB domain and a NPH3 domain, that is the domain that recognize the substrate. NPH3 domain

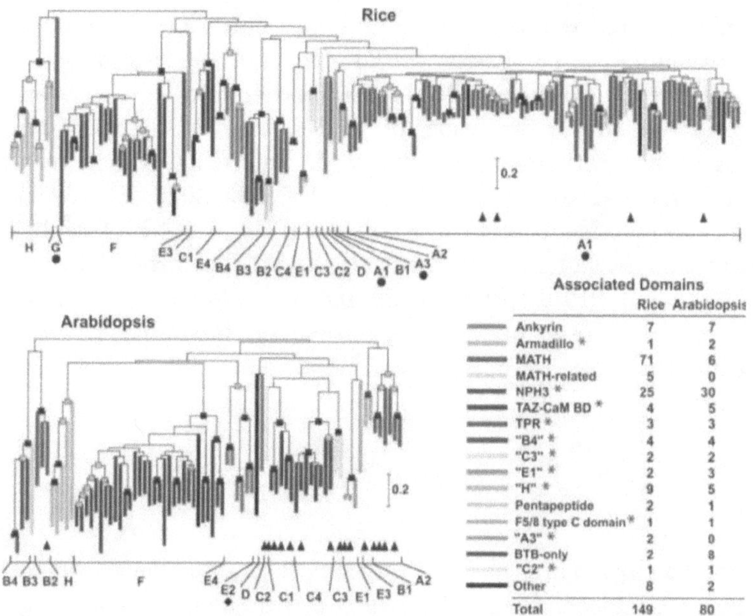

Figure 13. Phylogenetic trees of the complete BTB protein superfamilies in O. sativa and A. thaliana. The subfamilies are marked on the bottom right and colour-code by the nature of the domains either N- or C- terminal to the BTB domain. Closed circles indicate *O. sativa*-specific subfamilies. The closed diamond indicates an *A. thaliana*-specific E2 subfamily. Arrowheads indicate BTB protein shown to interact with Cul3. Reprinted from (Gingerich et al., 2007).

containing proteins are the most numerous BTB proteins in *A. thaliana* (**Figure 13**) and one of them is BTB/POZ PROTEIN HYPERSENSITIVE TO ABA 1 (BPH1), which is negatively involved in ABA signaling, however, its substrate targets have not been elucidated yet (Woo et al., 2018).

CRL3 E3 ligases are also supposed to be involved in defence against pathogens. Non-expresser of PR genes 1 (NPR1) is a BTB protein that plays an essential role in binding of SA. NPR1 oligomers become monomers under pathogen infection, and this process triggers SA accumulation (Zhang et al., 2019). In the absence of SA or pathogen challenge, NPR1 is degraded by the 26S proteasome and its inhibitory effect on effector-triggered cell death and anti-pathogen defence is eliminated. Interestingly, NPR1 does not interact directly with CUL3 (Mukhtar et al., 2009), therefore, it has been proposed that association of CUL3 with NPR1 may be mediated by other adaptors like NPR3 and NPR4, homologues of NPR1 that contain BTB domains (Fu et al., 2012). However, recent results questioned whether NPR3 and NPR4 really function as E3 ligases (Ding et al., 2018). They proposed that NPR1 and NPR3/NPR4 are SA receptors, but play opposite roles in transcriptional regulation of SA-induced gene expression.

3. Cullin 4-RING based E3 ligase (CRL4).

The last group are the CRL4^{DWD} complexes, which are formed by a CUL4 isoform, RBX1 and a substrate adaptor consisting of a DNA damage-binding protein (DDB)-1 and a set of DDB1-binding/WD-40 domain containing (DWD) proteins. DDB1 tether various DWD proteins to the CUL4 by two ß-propeller structures, one that connects to the N terminus of CUL4 and another that connects the various DWD proteins (Angers et al., 2006). There are 85 DWD proteins in *A. thaliana (Lee et al., 2008)*, including CONSTITUTIVELY PHOTOMORPHOGENIC (COP)-1 and the SUPPRESSOR OF PHYA (SPA)-1 protein family, that can work alone or together (Cardozo & Pagano, 2004); DE-ETIOLATED (DET)-1 (Bernhardt et al., 2006), the CDD complex containing DET1 and the enzymatically inactive E2 variant COP10 in association with DDB1 (Yanagawa et al., 2004).

CRL4^{DWD} complexes play a key role to regulate light response. COP1 is a RING protein that regulates responses to red/far red, blue and UV-B light (Lau & Deng, 2012). COP1 is involved in the rapidly degradation of several transcription factors in the dark. A notable substrate is HY5 (ELONGATED HYPOCOTYL (HY)-5 (Duek & Fankhauser, 2005). Red, far-red and blue light negatively regulate COP1, stabilizing these transcription factors upon light exposure, which then are able to promote photomorphogenesis. However, COP1 is able to interact with SUPPRESSOR OF PHY A-105 (SPA)-1-4 proteins, and the complex COP1-SPA associates with the core CUL4-RBX1 forming a CRL4^{DWD} E3 ligase (Callis, 2014). The role of the COP1 RING domain in this dual-RING complex is not clear, but in vitro ubiquitination of the transcription factor LONG AFTER FAR-RED LIGHT (LAF)-1 was enhanced by the SPA-COP1 complex compared to COP1 alone (Seo et al., 2003).

Finally, and as we previously described, ABA receptors are targeted for degradation at the nucleus by CRL4^{DDA1} (See above, **Regulation of ABA core components by post-transcriptional mechanisms and interacting proteins**).

4. Regulation of CRL complexes:

CRL function is very complex and requires many proteins working together at the correct time. First, the different substrate adaptors must identify appropriate targets among all the proteins in the cell and bring them closer to the correct CUL/RBX1 ligation machinery, which must be free from related adaptors. Then, ubiquitination happens rapidly via repetitive cycles of E2-Ub binding, either as monoubiquitination or polyubiquitination. Finally, when the target has been ubiquitinated enough, it must be released and the CRL complex disassembled to enable reuse of the CUL/RBX1 core by other awaiting adaptors (Hua & Vierstra, 2011). Given this complex scenario, it is understandable that the CRL abundance, assembly and activity are under a strong regulation. Some substrate adaptors are regulated transcriptionally and/or by microRNA- or exonuclease-mediated downregulation of their mRNA (An et al., 2010; Parry et al., 2009; Potuschak et al., 2006). It's also possible to control their activity by regulating the nuclear/cytoplasm partitioning of specific components (Spoel et al., 2009). Moreover, a substrate adaptor attached to a CRL without substrate can be ubiquitinated and be degraded (Bosu & Kipreos, 2008). This has been observed in *A. thaliana* by treating the plants with the proteasome inhibitor MG-132, which increases the abundance of FBX and BTB proteins (Hua & Vierstra, 2011). This auto-ubiquitination provides a broad-based mechanism to dampen the activity of individual CRLs when not needed without compromising the CUL/RBX1 core. Another mechanism is the dimerization of CRL3 (Lechner et al., 2011), that can theoretically increase their target range.

However, the most complex and possible the most influential regulatory mechanism affecting CRLs are two competing cycles (**Figure 14**). The first cycle involves RUB1 (Nedd8 in animals), which is conjugated to proteins via a three-step reaction cascade similar to the ubiquitin. The E3 activity appear to be conferred by the own RBX1 subunit of CRLs. This way, RUB1 is covalently attached to a conserved lysine of the CUL adjacent to the RBX1-binding site (Weber et al., 2005). This modification improves the affinity of the CUL/RBX1 core for the activated E2-Ub intermediate and induces a conformational change in the CUL, which tilts the E2 closer to the substrate-binding pocket (Duda et al., 2008). After the substrate is ubiquitinated enough and released, the binding of RUB1 can be reversed by the COP9/signalosome (CSN) complex. De-RUBylation is performed by CSN5 subunit, a zinc dependant metalloprotease (Lyapina et al., 2001). The second cycle of CRL regulation is driven by CULLIN-ASSOCIATED NEDD8-DISSOCIATED (CAND)-1 protein, which has the capacity to interact with all canonical CULs. This binding inhibits CRL activity by dissociating the adaptor module from the CUL/RBX1 core (Min et al., 2003). However, CAND1 binding cannot happen if the CUL is RUBylated. This way, CRLs are maintained in a dynamic equilibrium, which allows many competing adaptors to deliver substrates to be

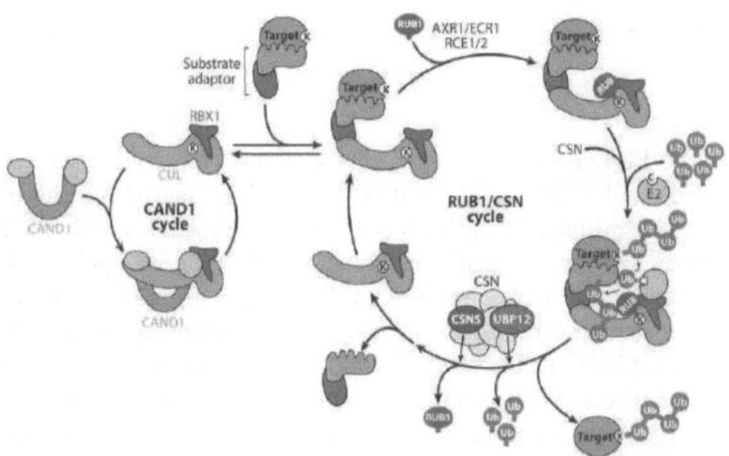

Figure 14. Proposed regulatory cycles of cullin-RING ligases (CRLs) involving CAND1 and RUB1-CSN-mediated RUBylation/de-RUBylation of the cullin (CUL) subunit. (a) Diagrams of the regulatory cycles. Via a transient and reversible binding of CAND1 to the CUL/RBX1 catalytic core, a dynamic pool of uncommitted CRLs is maintained in the cell. Occlusion of both the adaptor-binding and RUB1-binding sites by the U-shaped CAND1 prevents CRL assembly. Various adaptors then identify cellular substrates and recruit them to unsequestered CUL/RBX1 cores to generate an active CRL complex that enters the RUB1/CSN cycle. Through the action of the AXR1/ECR1 E1 heterodimer, the RCE1/2 E2, and an E3 activity provided in part by RBX1, RUB1 is attached to a positionally conserved lysine (K) at the C-terminal end of the CUL, which in turn helps activate the CUL/RBX1 ubiquitin (Ub) ligase activity. Ubs are subsequently added to the substrate and sometimes to the substrate adaptor (auto-ubiquitination), especially if no substrate is present. The ubiquitinated substrate is released and often degraded by the 26S proteasome. Either before or after the substrate is fully ubiquitinated, the engaged CRL complex associates with the eight-subunit COP9/signalosome (CSN). Through the action of the de-RUBylating activity provided by the CSN5 subunit and the deubiquitylating enzyme (DUB) activities provided in part by the associated DUB UBP12, RUB1 and Ubs bound to the CUL and the adaptor, respectively, are released from the CRL. Final dissociation of the substrate adaptor then allows the CUL/RBX1 core to re-enter the CAND1 and RUB1/CSN cycles for eventual reuse. Reprinted from (Hua & Vierstra, 2011).

ubiquitinated. At the same time, encourages sufficient engagement with the substrate as poly-ubiquitination proceeds but discouraging and/or reversing auto-ubiquitination (Hua & Vierstra, 2011).

BTB/POZ-MATH (BPM): substrate adaptors of CRL3^{BTB} ubiquitin ligase.

CRL3 E3 ligases are composed of a CUL3 protein that serves as scaffold. CUL3 binds in its C-terminal region RBX1 and in its N-terminal region, a protein containing a BTB fold. BTB proteins comprise a diverse group with 80 members in *A. thaliana* and 149 members in *O. sativa* (Gingerich et al., 2007), which have been divided into 12 subgroups based on their additional domains. While the BTB fold interacts with CUL3, the second domain functions as substrate adaptor for specific recognition of the substrate (Weber et al., 2005). One such domain is the MATH domain (**Figure 13**).

The BTB/POZ-MATH (BPM) family is formed by proteins which contain a BTB fold in their C-terminal and a MATH domain located within the first 200 amino acids of their N-terminal region (Chen et al., 2013). Interestingly, this family has largely expanded and diversified in some plant species, while there a 6 *BPM*s in *A. thaliana* (**Figure 15**), there are 54 members in *O. sativa* (Kushwaha et al., 2016). Additionally, there are other proteins with a MATH domain, which can be the single domain of the protein or be repeated two, three or four times (**Table 1**).

The MATH domain functions as the recognition site for various substrates of the CUL3^{BPM} E3 ligase. Some of these

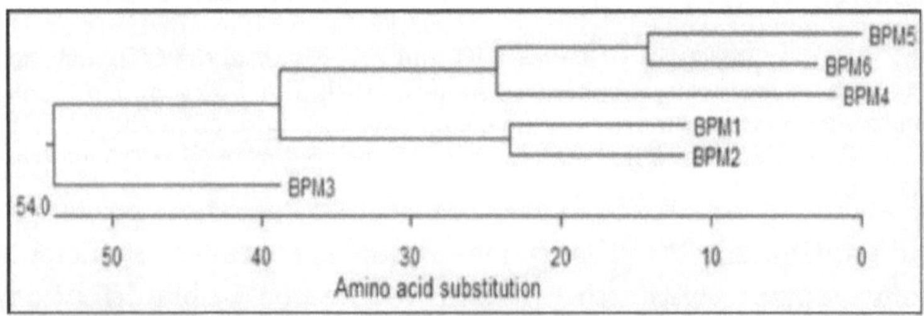

Figure 15. Cladogram of *A. thaliana* BPMs. Phylogenetic tree of the multiple sequence alignment of *A. thaliana* BPM family.

	Arabidopsis		Rice	
	Gene	Protein	Gene	Protein
Single MATH domain	28	39	13	15
Single MATH + Single BTB domain	06	09	54	57
Two MATH domain	25	31	01	01
Two MATH + Two BTB domain	–	–	01	01
Three MATH domain	02	02	–	–
Four MATH domain	01	01	–	–

Table 1. Comparison of MATH domain containing proteins and their encoding genes in *A. thaliana* and *O. sativa*.

substrates are already known, but it is much better studied in humans. Studies performed with the MATH-BTB SPOP protein (BPM orthologue in humans) to determine MATH-substrate interactions have revealed that the MATH domain can recognize a consensus peptide sequence in different targets, e.g. the phosphatase Puc, transcription factor Ci and histone macroH2A (**Figure 16A**) (Zhuang et al., 2009). The structure of SPOP reveals a dimer with 2 molecules of CUL3, thereby generating an ubiquitin ligase containing two substrate-binding sites and two catalytic cores (**Figure 16B**). Additionally, the MATH domain is able to adopt multiple orientations relative to the BTB domain through a flexible linker. For this reason, the combination of dimerization and flexibility has many implications for substrate ubiquitination, allowing more diverse substrates to be engaged. Moreover, dimerization may enable the substrate ubiquitin chains to adopt a greater variety of orientations (Zhuang et al., 2009).

In *A. thaliana*, the MATH domain of BPM3 interacts with the leucine zipper (ZIP) domain of AtHB6, a transcription factor from the class I homeobox-ZIP that negatively regulates ABA responses and is a target of the clade A PP2C ABI1 (Lechner et al., 2011). Reducing CRL3[BPM] function enhances the ABA-insensitive phenotype of lines overexpressing AtHB6, which indicates that CRL3[BPM] function positively regulates ABA

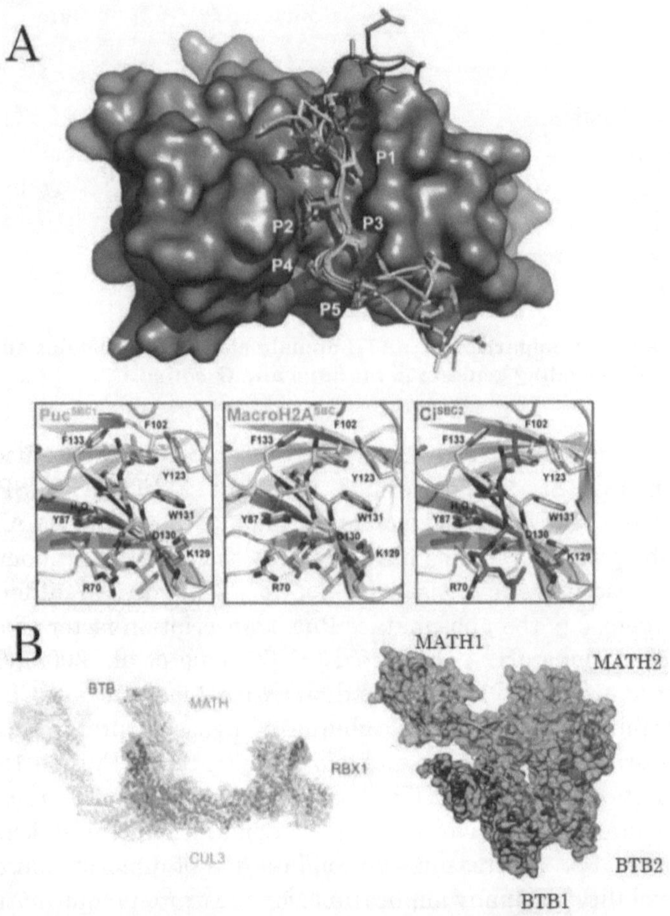

Figure 16. A putative model for the targeting of substrates by CRL3[BPM] complexes. (A) Comparison of 11 independent structures of isolated SPOP complexes with different substrates (3 from Puc, greens; 4 from MacroH2A, cyans/blues; 4 from Ci, pinks/magentas). After superposition over SPOP main chain, the substrates were displayed with backbones as cartoons and side chains as sticks, docked in the structure of SPOP (grey surface). The five amino acids that form the recognition motif (f-p-S-S/T-S/T) are indicated by P1–P5. Close-up views of SPOP (grey) complexes with Puc (green), MacroH2A (cyan), and Ci (magenta). Dashed lines, hydrogen bonds; red, oxygen; blue, nitrogen; red sphere, water. Reprinted from (Zhuang et al., 2009). (B) Left, atomic model of dimeric CUL3-RBX1-SPOP (PDB accession codes 1LDK, 4EOZ and 3HU6). Right, 3D structure prediction of BPM5 was performed over SPOP structure using Phyre2 (http://www.sbg.bio.ic.ac.uk/phyre2).

signaling. Recently, the DREB2A transcription factor, a key factor mediating transcriptional response to drought and heat stresses, has been identified as another target of CRL3[BPM], specifically BPM2 and BPM4 (Morimoto et al., 2017). Other transcription factors recognized by the BPMs are WRI1 (Chen et al., 2013) and MYB56 (Chen et al., 2015), involved in the control of fatty acid metabolism and flowering, respectively. It seems that various transcription factors present BPM recognition motifs, which implies a wide role of CRL3[BPM]-dependent proteolysis in transcriptional regulation.

OBJECTIVES

The main objective of this book was the study of the molecular mechanisms involved in clade A PP2C degradation and their contribution to ABA signaling. Specifically, we focused in the role played by the monomeric RING-type RGLG1 E3 ligase and the BPM3/5 substrate adaptors of multimeric CRL3 E3 ligases. Specific objectives are:

1. Characterization of the BPM3/5 substrate adaptors of CRL3^{BPM} complexes and their function as negative regulators of clade A PP2Cs.

   1.1. Study of the interaction among BPM3/5 and PP2Cs.

   1.2. Analysis of PP2Cs degradation and possible regulation by ABA.

   1.3. Generation and characterization of BPM3/5 OE lines and loss-of-function *bpm3 bpm5* double mutant plants to provide genetic evidence of their role in ABA signaling.

   1.4. Study of the ubiquitination of PP2CA in mutants impaired in BPMs.

2. Characterization of the molecular mechanism whereby ABA enhances PP2CA degradation by RGLG1.

   2.1. Analysis of the subcellular localization of RGLG1 and a RGLG1^{G2A} mutant.

   2.2. Study of the effect of ABA on RGLG1 and RGLG^{G2A} localization.

   2.3. Effect of abiotic stress on PP2CA degradation by RGLG1.

   2.4. Analysis of the formation of tertiary complexes among PP2CA, RGLG1 and ABA receptors in the presence of ABA.

RESULTS I: BPMs

BPM3 and BPM5 subunits of Cullin3-RING E3 ubiquitin ligase target clade A PP2Cs for degradation.

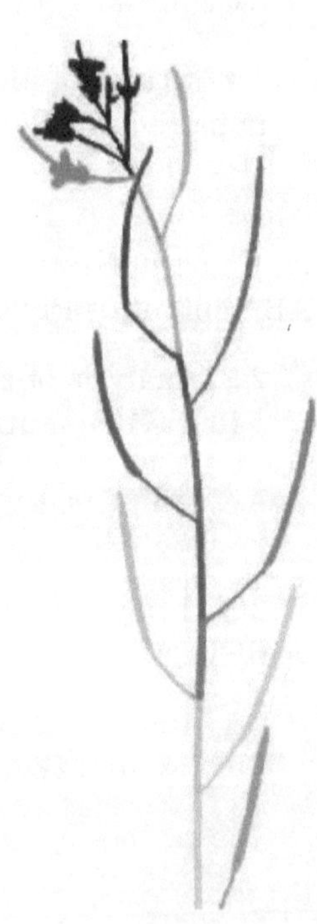

BPM3 and BPM5 interact with clade A PP2Cs.

Firstly, co-immunoprecipitation (coIP) coupled with liquid chromatography-tandem mass spectrometry (LC-MS/MS) was performed to identify proteins that co-immunoprecipitated *in vivo* with FLAG-tagged PP2CA expressed in *A. thaliana* (*35S$_{pro}$:3xFLAG-PP2CA*). Protein extracts were obtained from 2 weeks (w) old seedlings after different conditions, which included different incubation times of the proteasome inhibitor MG-132 and ABA. We found that native BPM3, BPM4 and BPM5 proteins co-immunoprecipitated with FLAG-PP2CA after 24 hours (h) treatment with MG-132 and 6h with ABA (**Table 2** and **Appendix Table S1**). MG-132 was sufficient to see interaction with BPM5, but not with BPM3 and BPM4. The additional identification of RGLG1

Condition	Identified proteins		Protein mass	Protein	Protein	p value
	Locus	Name	(Da)	score	seq. sig.	
Mock	AT3G11410	PP2CA	44235	164	8	0,00017
MG132 + ABA 12h	AT3G11410	PP2CA	44235	285	11	0,00018
	AT2G39760	BPM3	45318	42	1	0,00053
MG132 24h	AT3G11410	PP2CA	44235	332	13	0,000002
	AT5G21010	BPM5	45562	53	1	0,000015
MG132 24h + ABA 6h (1)	AT3G11410	PP2CA	44235	391	19	0,000012
	AT5G21010	BPM5	45562	67	1	0,0000006
MG132 24h + ABA 6h (2)	AT3G11410	PP2CA	44235	327	14	0,000026
	AT5G21010	BPM5	45562	168	3	0,00032
	AT3G03740	BPM4	51422	79	3	0,092
	AT2G39760	BPM3	45318	48	2	0,0008
	AT3G01650	RGLG1	53848	95	5	0,026

Table 2. Number of PP2CA, BPMs and RGLG1 peptides identified by LC-MS/MS in five independent experiments. Total proteins were extracted from 2w old *A. thaliana* seedlings (*35S$_{pro}$:3×FLAG-PP2CA*) treated without (mock) or with different mixes and times of 50µM MG-132 and 50µM ABA. Immunoprecipitation (IP) was performed using α-FLAG and the IP products were analysed by LC-MS/MS.

during this analysis validates this screening system, as it has been reported as a PP2CA interacting protein (Wu et al., 2016).

We concentrated further work on BPM5 as a result of recovering the highest number of peptides from it (**Table 2** and **Appendix Table S1**). Also on BPM3 because it showed the highest expression at the AtGenExpress public database (**Figure 17**).

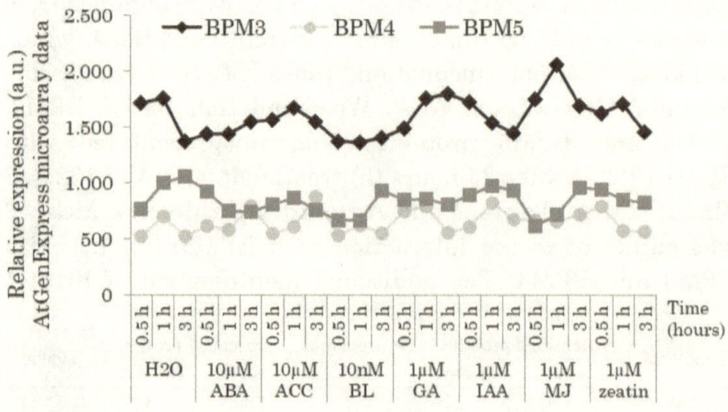

Figure 17. Relative expression of BPM3, BPM4 and BPM5 in seedlings. Seedlings of *A. thaliana* were treated without (H2O) or with different hormones: 10mM ABA, 10mM ACC, 10mM BL, 10mM GA, 10mM IAA, 10mM MJ or 10mM zeatin. Expression data was obtained from Affymetrix microarrays for *A. thaliana* deposited into the AtGenExpress public database (Goda et al., 2008), visualized using the AtGenExpress Visualization Tool located in http://weigelworld.org/resources.html.

BPM3 and BPM5 are located in the nucleus (**Figure 18**), as the PP2CA (Antoni et al., 2012). But also as other PP2Cs of the clade A as HAB1, ABI1 and ABI2 (Mitula et al., 2015; Saez et al., 2008), for that reason, the next step was to analyse the interaction of BPM3 and BPM5 with all this PP2Cs. We performed bimolecular fluorescence complementation (BiFC) assays and found that PP2CA, HAB1, ABI1 and ABI2, showed

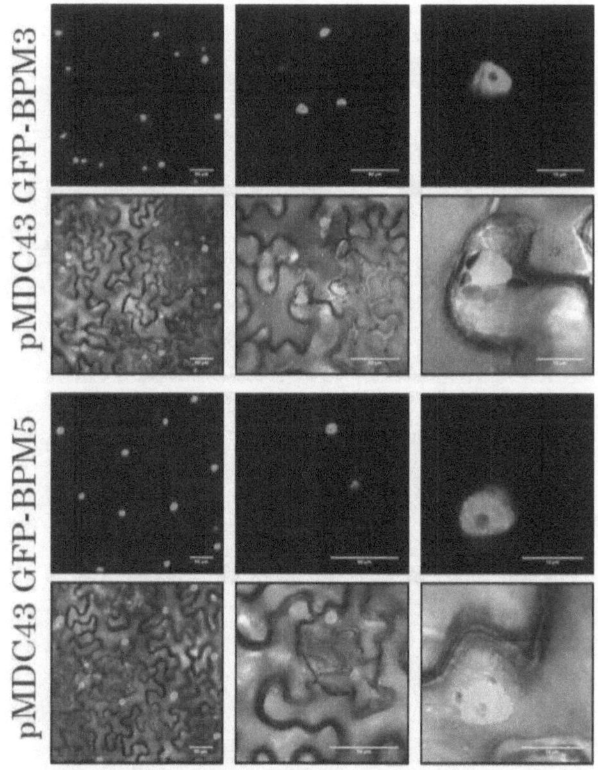

Figure 18. BPM3 and BPM5 proteins are localized in the nucleus of *Nicotiana benthamiana* leaf cells. Confocal images showing the nuclear localization of GFP-BPM3 or GFP-BPM5 fusion proteins in transiently transformed *N. benthamiana* leaf cells. Plants were infiltrated with *Agrobacterium tumefaciens* harbouring the indicated construct. Scale bars = 50 μm (general view) or 10/15 μm (zoom).

nuclear interaction with both BPM3 and BPM5 (**Figure 19**). The resultant nuclear speckles were different in shape. On the contrary, and as negative controls for the experiment, BPM3 and BPM5 did not interact with PYL8, while the PP2Cs did not interact with OST1$^{\Delta 280}$ (Vlad et al., 2009).

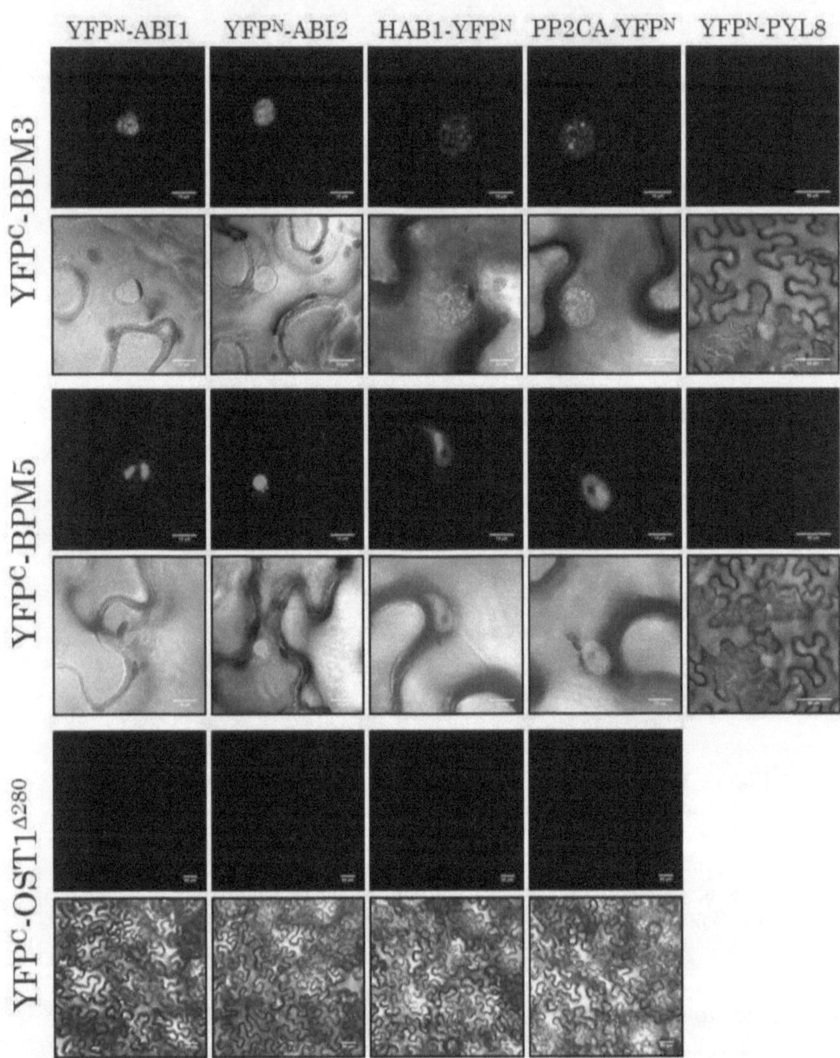

Figure 19. BiFC assays show nuclear interactions of BPM3 or BPM5 with ABI1, ABI2, HAB1 or PP2CA in *N. benthamiana* leaf cells. Confocal images of transiently transformed *N. benthamiana* leaf cells co-expressing either YFP^C-BPM3 or YPF^C-BPM5 and the indicated YFP^N-PP2C protein. As a negative control for the YFP^C-BPMs, interaction with YFP^N-PYL8 was not observed. For the PP2Cs we used YFP^C-OST1$^{\Delta280}$, which has already been publish (Vlad et al., 2009). Scale bars = 50μm (general view) or 10μm (zoom).

BPMs are composed of two domains: MATH and BTB (**Figure 20**). The MATH domain contain the substrate recognition module, so we tested whether it might recognize PP2Cs. To this end we split BPM3 in two regions, the N-terminal

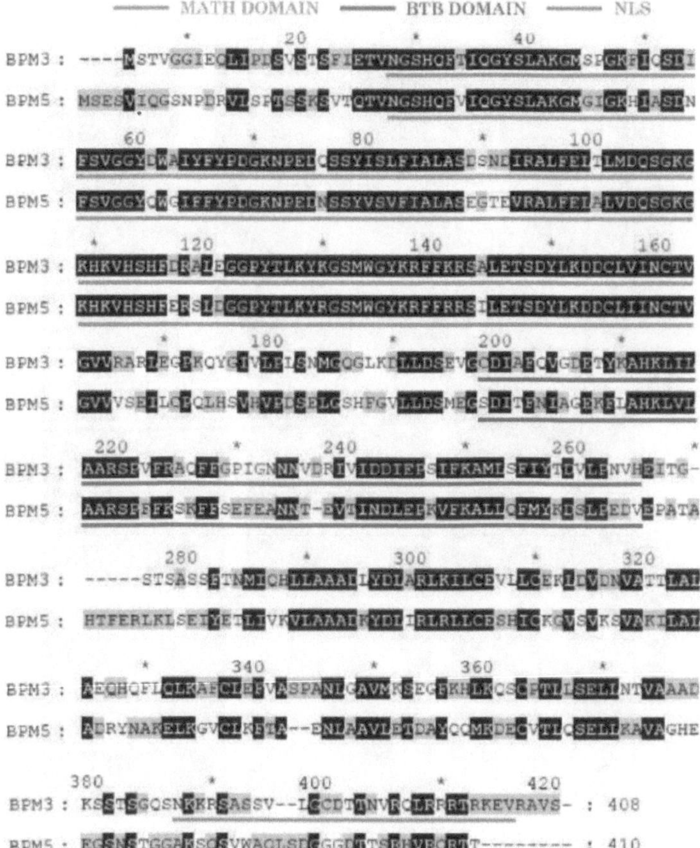

Figure 20. Amino acid alignment of BPM3 and BPM5. The proteins were aligned using the CLUSTAL OMEGA method with MegAlign Pro (15.0.0). The domains are indicated with different colour lines. The MATH (blue) and BTB (red) domains were predicted using ScanProsite located in https://prosite.expasy.org/ (de Castro et al., 2006). The nuclear localization site (NLS) is indicated with a green line and was predicted using cNLSmapper located in http://nls-mapper.iab.keio.ac.jp/cgi-bin/NLS_Mapper_form.cgi (Kosugi et al., 2009).

containing the MATH domain (residues 1-175) and the C-terminal containing the BTB domain (residues 176-408). As expected, only the MATH region of BPM3 showed interaction with PP2CA (**Figure 21A**).

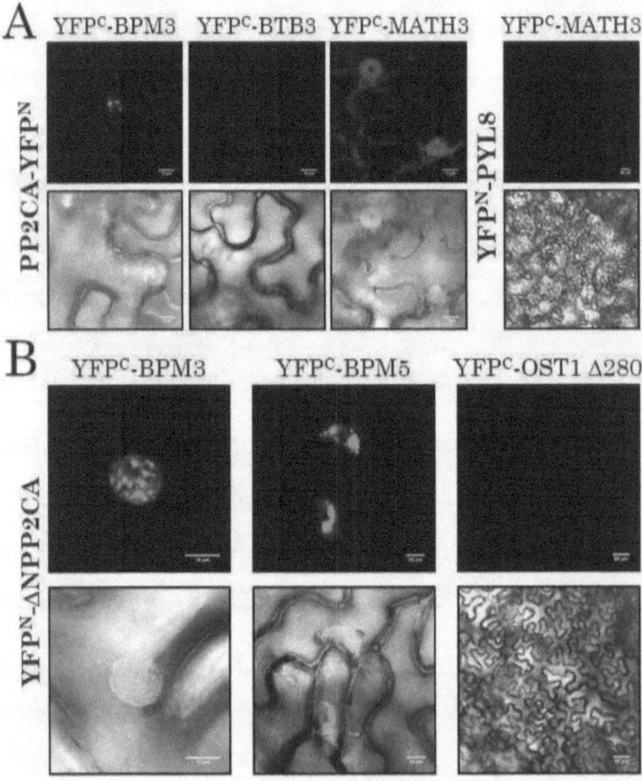

Figure 21. The MATH domain of BPM3 and the PP2C catalytic core are sufficient for the nuclear interaction of PP2CA with BPM3 and BPM5. Confocal images of transiently transformed *N. benthamiana* leaf cells co-expressing (A) PP2CA-YFP^N or YFP^N-PYL8 and YFP^C-BPM3 or the individual BTB or MATH modules of BPM3, indicated as BTB3 and MATH3, respectively; or (B) YFP^N-ΔNPP2CA and YFP^C-BPM3, YFP^C-BPM5 or YFP^C-OST1^Δ280. Scale bars = 50μm (general view) or 10μm (zoom).

The interaction between the MATH region of BPM3 and the PP2CA is nuclear and cytosolic, instead of only nuclear, probably as a result of the NLS being on the C-terminus of BPM3, which is in the BTB region (**Figure 20**). The non-interaction with PYL8 was used as a negative control for the MATH.

We also tested whether the N-terminus region of PP2CA was required for the interaction with the BPMs. This region corresponds to the most variable part on all clade A PP2Cs. ΔNPP2CA (lacking amino acid residues 1-99) was assayed by BiFC. We found that the catalytic core of PP2CA was sufficient for the interaction with both BPM3 and BPM5 (**Figure 21B**). As with the other PP2Cs, the non-interaction with OST1$^{\Delta 280}$ was used as a negative control for the ΔNPP2CA.

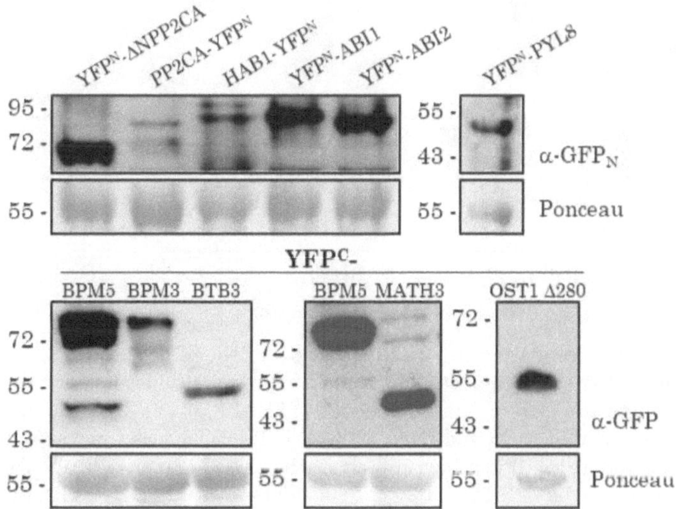

Figure 22. Immunoblot analysis of the YFP^N- and YFP^C-fusion proteins used in the BiFC experiments above. All proteins extracted from the BiFC samples shown expression. Ponceau staining of Rubisco was used as a protein loading control.

Eventually, we also verified by immunoblot that the YFP^N- and YFP^C-fusion proteins were correctly expressed (**Figure 22**). Additionally, the reconstituted YFP signal in the BiFC experiments described above was quantified (**Figure 23**).

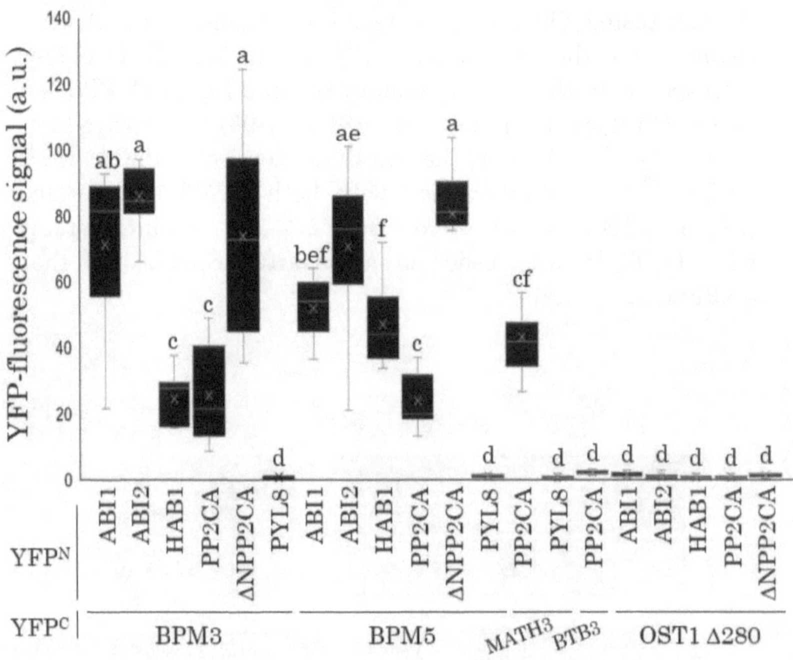

Figure 23. Box plot showing quantification of the YFP signal recovered in different BiFC experiments. For each interaction 10 independent nuclei were analysed with the help of ImageJ software. YFP^C-BPM3, YFP^C-BPM5 and YFP^C-MATH3 interact with the indicated PP2Cs. YFP^N-PYL8 was used as a negative control for the YFP^C-BPM fusion proteins, whereas YFP^C-OST1$^{\Delta280}$ for the YFP^N-PP2C fusion proteins. YFP^C-BTB3 does not interact with PP2CA-YFP^N. Median is indicated with a line, mean with an x and the letters represent different groups formed after performing a One-way ANOVA analysis with post-hoc Tukey test of the data, with a significance level of 0.05.

After that, we performed pull-down assays with His-tagged
PP2CA and the others clade A PP2Cs (HAB1, ABI1 and
ABI2). We used GST-BPM5 as bait. GST-BPM3 was not used
because is poorly soluble and tends to unspecific precipitation.
We found once again that all the PP2Cs tested interacted with
BPM5 (**Figure 24**). However, there was not interaction of the
PP2Cs with GST. For that reason, we used it as a negative
control through all the pull-down experiments. GST-PP2CA
also interacted with the His-MATH domains of BPM3 and
BPM5 (**Figure 25A**). Moreover, His-ΔNPP2CA interacted
with GST-BPM5 (**Figure 25B**). With both experiments we
confirmed that the interaction between the PP2CA and the
BPM3 or BPM5 is happening through the catalytic core of the
PP2CA with the MATH domain of the BPMs.

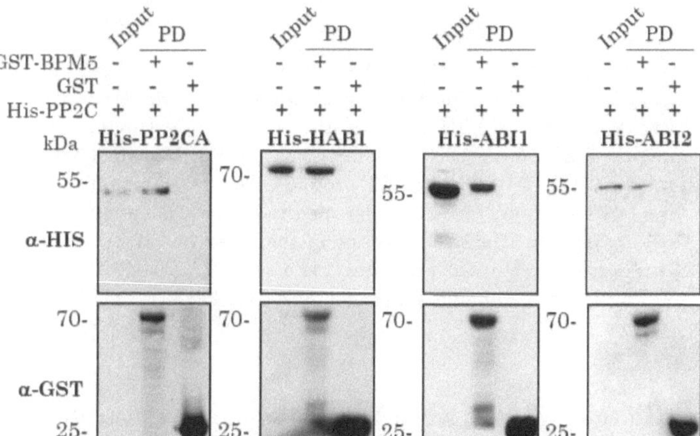

Figure 24. Pull-down (PD) assay showed interaction of PP2CA,
HAB1, ABI1 or ABI2 with BPM5. The indicated His-tagged PP2Cs were
incubated with immobilized GST-BPM5 or GST proteins. After washing,
proteins were eluted with Laemmli buffer and detected by immunoblot
analysis using α-His or α-GST antibodies. A fraction of the input and the
corresponding His-tagged PP2C recovered are indicated.

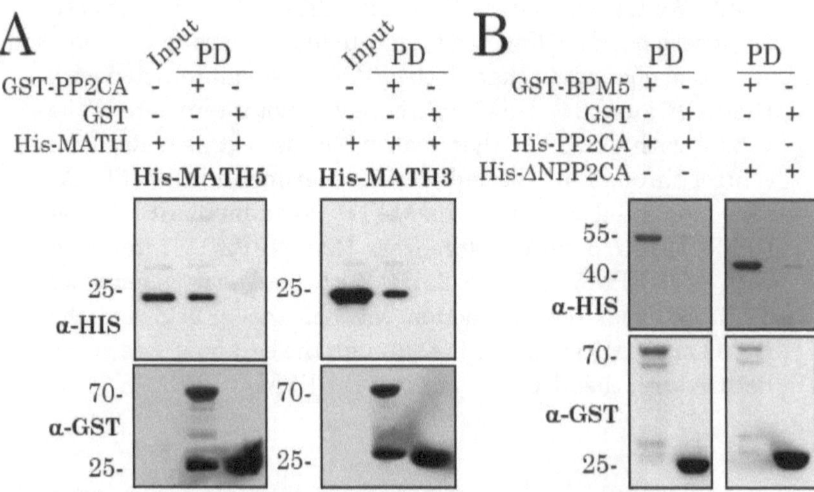

Figure 25. PD assay showed that the interaction between the PP2CA and BPM5 is happening through the catalytic core of the PP2CA and the MATH domain of the BPM. (A) His-tagged MATH domain of BPM5 (MATH5) or BPM3 (MATH3) were incubated with immobilized GST-PP2CA or GST proteins. After washing, proteins were eluted with Laemmli buffer and detected by immunoblot analysis using α-His or α-GST antibodies. A fraction of the input and the corresponding His-tagged MATH recovered are indicated. (B) His-PP2CA or His-ΔNPP2CA were incubated with immobilized GST-BPM5 or GST proteins.

In addition, we performed Y2H assays to demonstrated these interactions. We observed again the interaction of PP2CA, HAB1 or ABI1 fused to the Gal4 activation domain (AD) with either BPM3 or BPM5 fused to the Gal4 DNA binding domain (BD) (**Figure 26**). We used the empty domains as negative controls of the interaction, in which ones we can clearly see a strong difference in growth. Especially at the most diluted cultures.

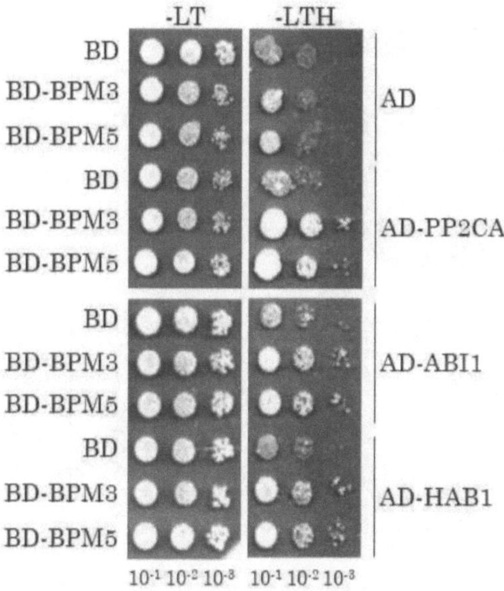

Figure 26. Y2H interaction of BPM3 or BPM5 with PP2CA, ABI1 or HAB1. Transformed yeast with BD, BD-BPM3 or BD-BPM5 and AD, AD-PP2CA, AD-ABI1 or AD-HAB1, was grown overnight in liquid synthetic (SD) medium lacking Leu and Trp. Dilutions of these cultures were dropped on either control medium lacking Leu and Trp (SD -LT) or selective medium additionally lacking His (SD-LTH). Yeasts were allowed to grow for 3 days (d) at 28°C before interaction was check off.

To further validate the interactions observed for PP2CA and HAB1 with the BPMs, we performed a split-luciferase (split-LUC) complementation assay in *N. benthamiana* leaves (**Figure 27A**). In contrast to BiFC assays, the protein-protein interaction is reversible, which makes it more reliable. In these experiments we used the empty vectors as negative controls. Addinionatly, another set of *N. benthamiana* leaves were used to quantify this interaction. For this we also co-infiltrated GUS to use it as a normalizer of the signal (**Figure 27B**).

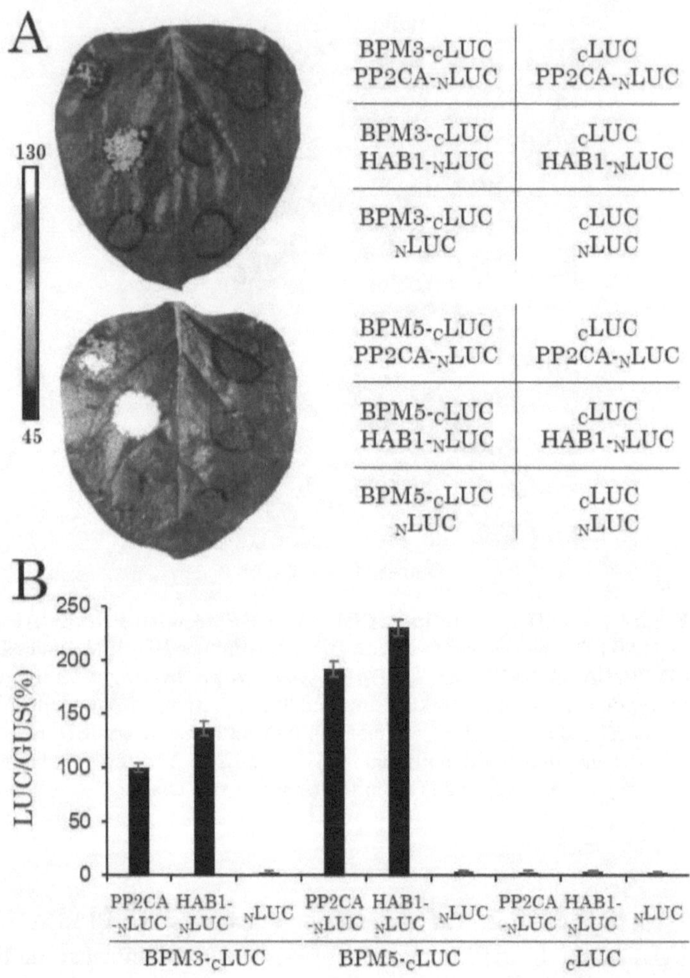

Figure 27. Split-LUC complementation assay reveals interaction of BPM3 or BPM5 with PP2CA or HAB1. (A) BPM3-cLUC, BPM5-cLUC or cLUC were pair with PP2CA-nLUC, HAB1-nLUC or nLUC. The indicated constructs were co-expressed in *N. benthamiana* leaves (showed in grey) by *A. tumefaciens* mediated infiltration. After application of 1mM D-luciferin the luciferase (LUC) activity was screened with a CCD system. LUC signal was converted to false colours with ImageJ software. Colour scale represent LUC activity. (B) *N. benthamiana* leaves were infiltrated with the same constructs but also co-expressing GUS. LUC activity was quantified from protein extracts by a luminometer. While GUS activity was also quantified by fluorometry and used to normalize the LUC signal. Data are means of two independent experiments ± SD.

Finally, we performed coIP assays of PP2CA-GFP, GFP-ABI1 or GFP-HAB1 and either HA-BPM3 or HA-BPM5. The different partners were co-expressed in *N. benthamiana* leaves by *A. tumefaciens* mediated infiltration. These assays revealed that HA-BPM3 or HA-BPM5 co-immunoprecipitated with the immobilized GFP-tagged PP2C (**Figure 28**). As a negative control we used immobilized GFP alone. With which the BPMs did not co-immunoprecipitated.

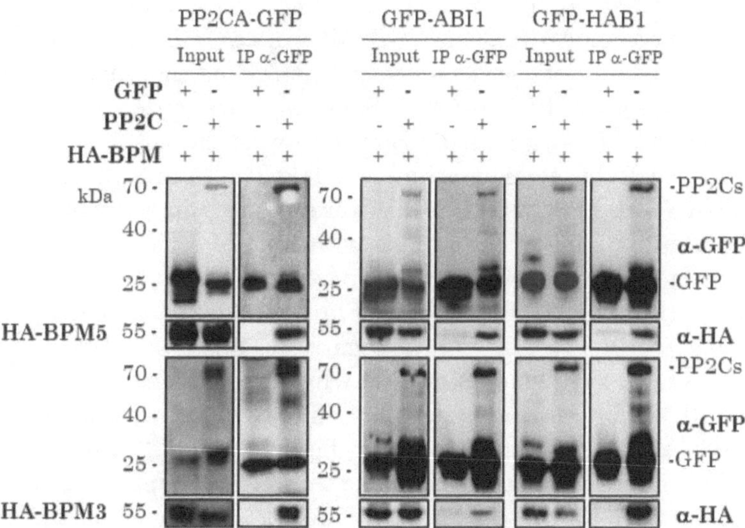

Figure 28. Coimmunoprecipitation of HA-BPM5 or HA-BPM3 with immobilized PP2CA-GFP, GFP-ABI1 or GFP-HAB1.** The pairs were co-expressed in *N. benthamiana* leaves by *A. tumefaciens* mediated infiltration. Immunoprecipitation (IP) of the proteins extracts was done with α-GFP antibodies. Each immunoprecipitated PP2C was probed with α-HA antibodies to detect coIP of either HA-BPM5 (top panels) or HA-BPM3 (bottom panels). As a negative control of each experiment, IP of GFP alone was performed. But in this case neither BPM coimmunoprecipitated.

BPM3 and BPM5 promote degradation of clade A PP2Cs *in vivo*.

With the interaction of BPM3 and BPM5 with PP2CA, HAB1, ABI1 and ABI2 already confirmed. The next step was to study if the BPMs promote degradation of clade A PP2Cs. Firstly, we performed co-infiltration experiments in *N. benthamiana* leaves by *A. tumefaciens* mediated infiltration. Increasing amounts of the *A. tumefaciens* that drives expression of HA-BPM3 or HA-BPM5 were co-infiltrated with a fixed amount of the *A. tumefaciens* encoding the construct that expresses PP2CA-GFP. Samples were collected for detection of the expressed proteins by western blot and the RNA levels by semiquantitative RT-PCR (**Figure 29** and **Figure 30**). PP2CA-GFP levels were quantify in relation to the total proteins in the sample with ImageJ software. This data is shown in a histogram. Increasing amounts of HA-BPM3 (**Figure 29**) or HA-BPM5 (**Figure 30**) led to decreasing levels of PP2CA-GFP. As a negative control we also co-expressed a fixed amount of RFP, which was not degraded by the BPMs. Interestingly, the addition of ABA at the 4:1 ratio (PP2CA:BPM) enhanced the degradation of the PP2CA-GFP. This is not cause by the ABA alone cause is not happening at the 4:0 ratio. At the 4:4 ratio this effect is still shown for the BPM3. But for the BPM5 all the PP2CA-GFP was already degraded without ABA. Seeing the RNA levels, we could confirm that it was actual degradation of the PP2CA-GFP. Cause the transcription was not affected. Then, we performed similar experiments co-expressing fixed amounts of GFP-ABI1 or GFP-HAB1 with increasing amounts of HA-BPM3 or HA-BPM5. These increasing amounts also led to decreasing levels of GFP-ABI1 and GFP-HAB1 (**Figure 31**).

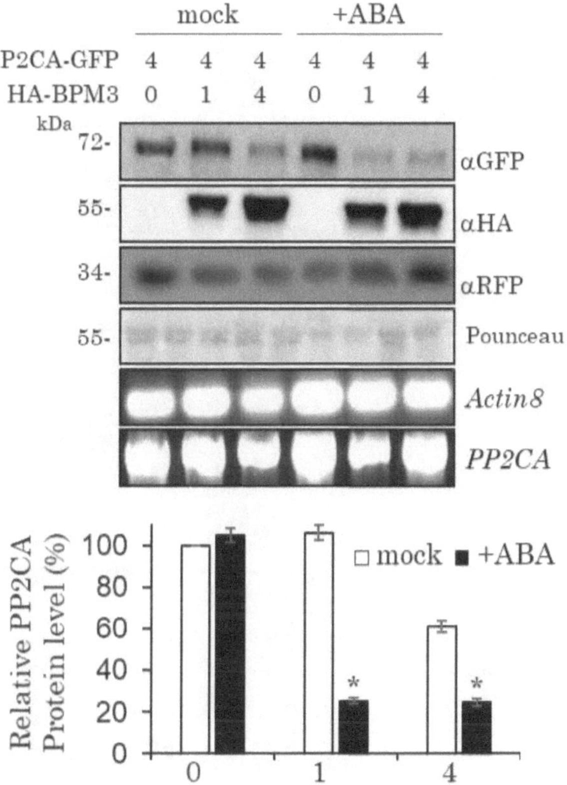

Figure 29. *In vivo* **degradation of PP2CA was observed in agroinfiltration experiments with increasing amounts of BPM3 in** *N. benthamiana*. Increasing amounts of an *A. tumefaciens* that drives the expression of HA-BPM3 were infiltrated with a constant amount of another agrobacteria driving the expression of PP2CA-GFP. The ratio of the relative concentrations of agrobacteria used in the different coinfiltrations are indicated by numbers (top). Cell extracts were analysed using α-HA to detect HA-BPM3, α-GFP to detect PP2CA-GFP and α-RFP to detect RFP. Which is not degraded by the BPM3. Ponceau staining of Rubisco was used as a protein loading control. mRNA expression levels of *PP2CA* and *ACTIN8* were analysed by semiquantitative RT-PCR. The histogram shows the quantification of HA-BPM3-promoted degradation of PP2CA-GFP. Data are means of two independent experiments ± SD. Asterisks indicate p < 0.05 (Student's t test) when comparing data obtained with or without 50µM ABA treatment.

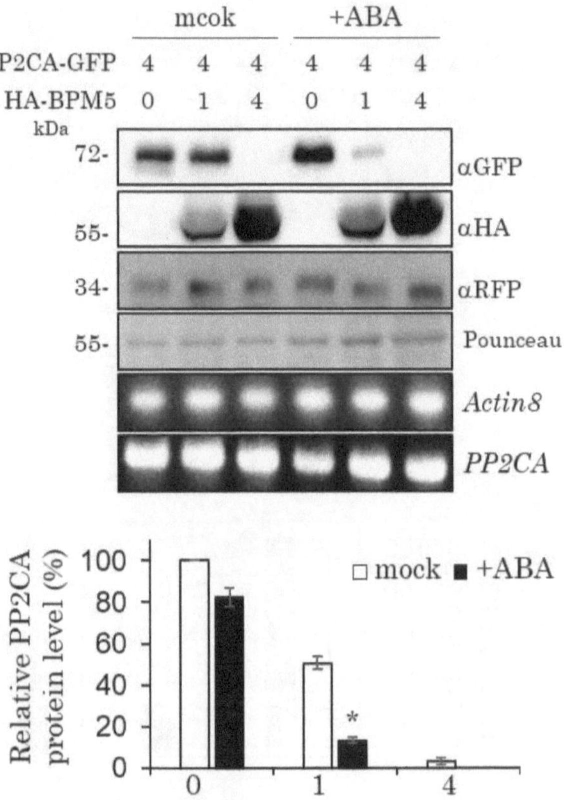

Figure 30. *In vivo* **degradation of PP2CA was observed in agroinfiltration experiments with increasing amounts of BPM5 in N. benthamiana.** Increasing amounts of an *A. tumefaciens* that drives the expression of HA-BPM5 were infiltrated with a constant amount of another agrobacteria driving the expression of PP2CA-GFP. The ratio of the relative concentrations of agrobacteria used in the different coinfiltrations are indicated by numbers (top). Cell extracts were analysed using α-HA to detect HA-BPM5, α-GFP to detect PP2CA-GFP and α-RFP to detect RFP. Which is not degraded by the BPM5. Ponceau staining of Rubisco was used as a protein loading control. mRNA expression levels of *PP2CA* and *ACTIN8* were analysed by semiquantitative RT-PCR. The histogram shows the quantification of HA-BPM5-promoted degradation of PP2CA-GFP. Data are means of two independent experiments ± SD. The asterisk indicates p < 0.05 (Student's t test) when comparing data obtained with or without 50µM ABA treatment.

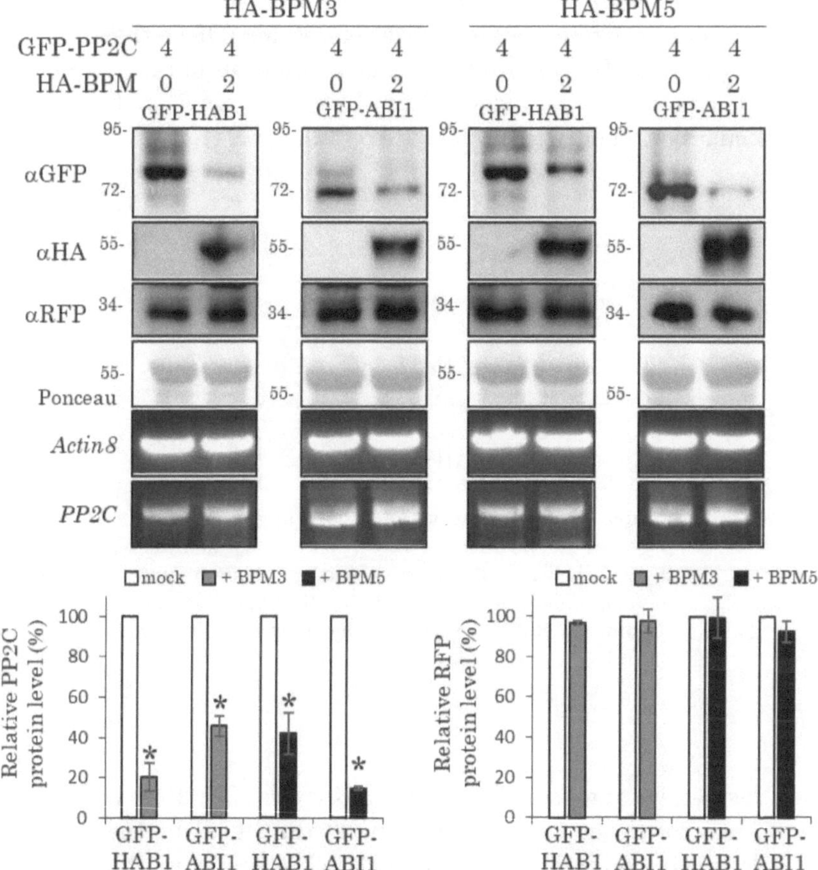

Figure 31. *In vivo* degradation of HAB1 and ABI1 was observed in agroinfiltration experiments with increasing amounts of BPM3/5 in *N. benthamiana*. An *A. tumefaciens* that drives the expression of HA-BPM3 or HA-BPM5 was infiltrated with a constant amount of another agrobacteria driving the expression of GFP-HAB1 or GFP-ABI1. The ratio of the relative concentrations of agrobacteria used in the different co-infiltrations are indicated by numbers (top). Cell extracts were analysed using α-HA to detect HA-BPM3/5, α-GFP to detect GFP- tagged HAB1 or ABI1 and α-RFP to detect RFP. Ponceau staining of Rubisco was used as a protein loading control. mRNA expression levels of *HAB1*, *ABI1* and *ACTIN8* were analysed by semiquantitative RT-PCR. The histogram shows the quantification of HA-BPM5-promoted degradation of PP2CA-GFP (left), whereas RFP was not affected (right). Data are means of two independent experiments ± SD. Asterisks indicate p < 0.05 (Student's t test) when comparing plants treated with or without HA-tagged BPMs.

From this point on, we wanted to work on stable lines of *A. thaliana*. Our first ones were overexpressing (OE) either HA-BPM3 or HA-BPM5. After immunoblot analysis of the protein extracts of three independent lines, we saw a higher accumulation of HA-BPM5 (**Figure 32**).

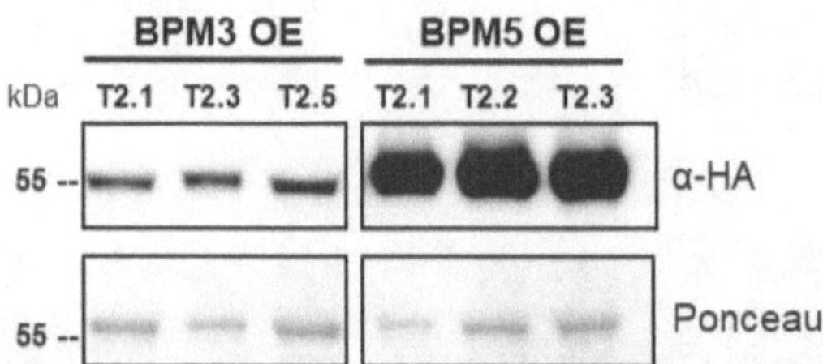

Figure 32. Overexpression of HA-tagged BPM3/5 proteins in *A. thaliana* lines. Immunoblot analysis confirm the expression of HA-tagged BPM3/5 proteins in *A. thaliana* BPM3 and BPM5 OE lines. Ponceau staining of Rubisco was used as a protein loading control.

Besides, we analysed the endogenous levels of PP2CA. We did it on root tissue to avoid masking of the signal by the very close rubisco large subunit (55 kDa). The first thing to point out is that this experiments are easier to quantify in presence of ABA thanks to the rapid induction of clade A PP2Cs (Wang et al., 2019b). We considered this as a regulatory mechanism to avoid the ABA signaling lasts for too long. By immunoblot analysis, using an α-PP2CA antibody, we found lower endogenous PP2CA levels in the OE lines (**Figure 33**). This was apparent in mock and ABA treated samples.

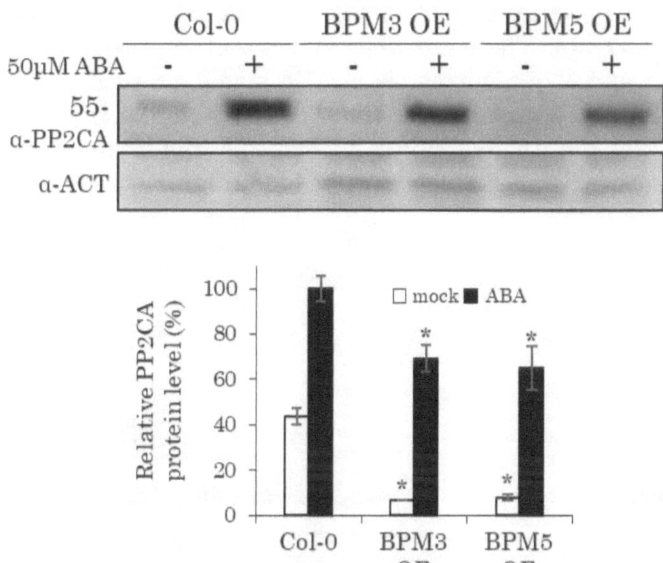

Figure 33. BPM3 and BPM5 promote degradation of PP2CA in *A. thaliana*. PP2CA protein levels are higher in Col-0 plants compared to BPM3 or BPM5 OE lines. This was analysed in mock- or 50µM ABA-treated plants. The ABA treatment was performed for 3h. Root protein extracts were analysed by immunoblot using α-PP2CA (α-E2663) to detect endogenous PP2CA protein levels. Actin (ACT) was analysed as a loading control with a specific antibody. Data are means of three independent experiments ± SD. Asterisks indicate p < 0.05 (Student's t test) when comparing data of BPM3 and BPM5 OE lines to Col-0 under the same conditions.

The next step was the generation of a *bpm3 bpm5* loss-of-function line that didn't express endogenous *BPM3* or *BPM5* (**Figure 34**). Conversely to the OE lines, we observed a higher accumulation of PP2CA compared to the wild type after ABA treatment by immunoblot analysis using α-PP2CA antibodies (**Figure 35**). There are other PP2Cs that are also induced by ABA and play a major role in ABA signaling, such as HAB1 and ABI1. Both PP2Cs accumulated more in the *bpm3 bpm5* mutant than in the wild type. We saw this by immunoblot analysis using specific antibodies α-HAB1 and α-ABI1 (**Figure 35**).

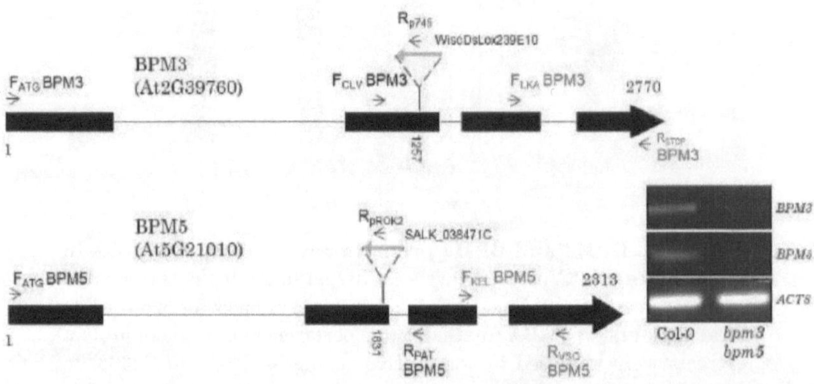

Figure 34. Structure of the *bpm3* and *bpm5* T-DNA insertion lines. Schematic diagram of the *BPM3* and *BPM5* genes showing the position of the T-DNA insertion in the *bpm3* and *bpm5* alleles. As well as the primers used for the genotype of the mutants (in black) and qRT-PCR analysis (in red). It was also analysed the relative expression of the *BPM3* and *BPM5* genes in the *bpm3 bpm5* mutant line compared to Col-0 wt.

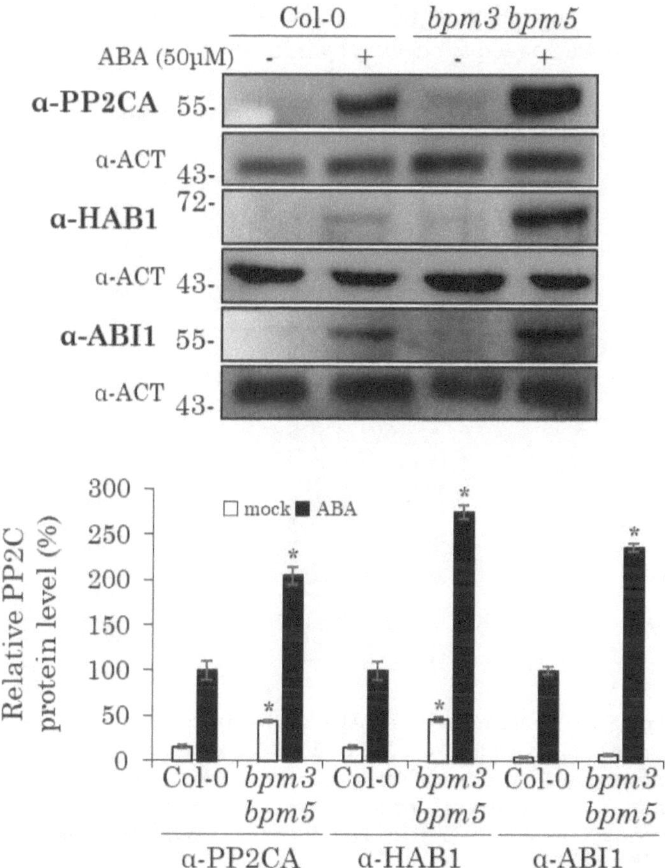

Figure 35. Enhanced accumulation of PP2CA, HAB1 and ABI1 in the *bpm3 bpm5* double mutant line compared to the Col-0 wild type. *A. thaliana* plants were treated for 3h with or without 50µM ABA. Root protein extracts were analysed by immunoblot using α-PP2CA, α-HAB1 and α-ABI1. Actin was analysed as a loading control with a specific antibody. Data are means of two independent experiments ± SD. Asterisks indicate p < 0.05 (Student's t test) when comparing data of *bpm3 bpm5* double mutant line to Col-0 under the same conditions.

We acknowledge Sean Cutler for give us the α-HAB1 antibody, which was not published yet. Therefore, before using it we needed to check it by immunoblot analysis (**Figure 36**).

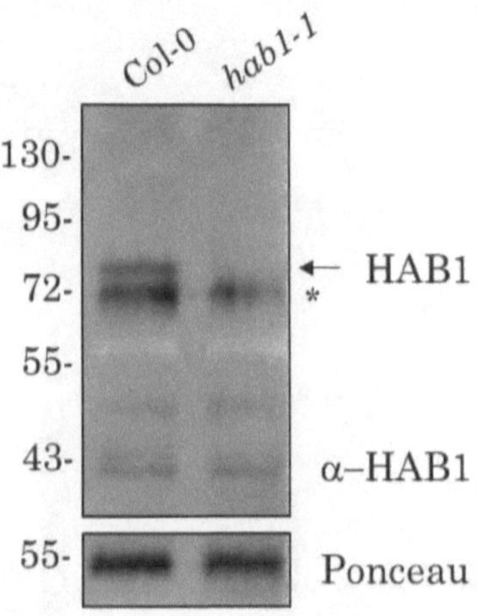

Figure 36. Specificity of α-HAB1 antibody. Seedlings of Col-0 and *hab1-1* were grown for 7d in MS medium. Total proteins were extracted and analysed by immunoblot using the α-HAB1 antibody. Ponceau staining of Rubisco was used as a protein loading control. The asterisk indicates unspecific recognition of a protein below the actual HAB1 (indicated by an arrow).

Finally, we performed a time course of the PP2CA in presence of cycloheximide (CHX). CHX is an inhibitor of protein synthesis in eukaryotes, which inhibit translation elongation through binding to the 60S ribosomal unit. This allowed us to differentiate if the BPMs were promoting the degradation of PP2CA, or blocking its synthesis. After an ABA treatment to accumulate PP2CA in the root tissue, we applied the CHX. We recovered material at 4 different times (0, 1, 3 and 6h) and tested the protein extracts by immunoblot analysis with α-PP2CA antibody. As a result, degradation of PP2CA was slower in the *bpm3 bpm5* plants compared to the wild type (**Figure 37**). However, after 6h of CHX treatment around 60% of the endogenous PP2CA had been degraded. This indicate that additional E3s (such as RGLG1/5) or non-26S proteasome pathways are also involved in PP2CA degradation (Wu et al., 2016; Yu & Xie, 2017).

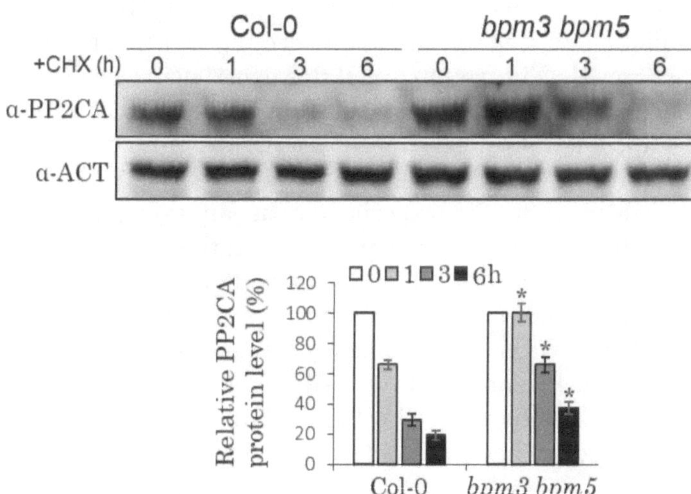

Figure 37. Degradation of PP2CA is delayed in the *bpm3 bpm5* double mutant line compared to the Col-0 wild type. Seedling of Col-0 or *bpm3 bpm5* were grown in liquid MS medium for 10d. Then, they were supplemented with 50µM ABA for 3h to induce PP2CA expression. After washing the ABA, 50µM CHX was added and root tissue was harvested at 0, 1, 3 and 6h. Actin was analysed as a loading control with a specific antibody. Data are means of two independent experiments ± SD. Asterisks indicate p < 0.05 (Student's t test) when comparing data of *bpm3 bpm5* double mutant line to Col-0 under the same conditions.

BPM3 and BPM5 gain-of-function leads to enhanced sensitivity to ABA.

To avoid the functional redundancy of the BPM genes we started our ABA-response phenotypical analysis using the OE lines described before (**Figure 32**). As stated above, these lines have a lower amount of PP2CA compared to Col-0 lines (**Figure 33**). Firstly, we found an enhanced ABA-mediated inhibition of seed germination, seedling establishment and root growth (**Figure 38**). In all this experiments we used two independent lines of each BPM: BPM3 2.3, BPM3 2.5, BPM5 2.1 and BPM5 2.3. The *hab1-1 abi1-2* loss-of-function line is an ABA-hypersensitive mutant (Saez et al., 2006). Therefore, we added it to the experiments as an example of high sensitivity to ABA. After 3d in the plates, the germination was scored. Both BPM5 OE lines had a similar high sensitivity to ABA as the *hab1-1 abi1-2* mutant line (**Figure 38A**). While, both BPM3 OE lines were slightly less sensible. 5d after that, the establishment was also scored. All four BPM OE lines had a similar sensitivity to ABA (**Figure 38B & 21D**). Around half of the seedling had establish compared to Col-0.

Simultaneously, a root growth experiment with the same lines was also performed. Once again, all four BPM OE lines had a similar growth arrest (**Figure 38C& 21E**). The roots of these lines grew approximately ~40% less than Col-0 roots in presence of ABA. Therefore, they were ABA hypersensitive compared with Col-0.

After that, we thought that if these lines presented a higher sensitivity to ABA, they could also show a water phenotype. We found that detached leaves of BPM3 and BPM5 OE lines showed reduced water loss compared to Col-0 (**Figure 39**). Col-0 leaves lost around 50% more water than the BPM OE lines under the same conditions.

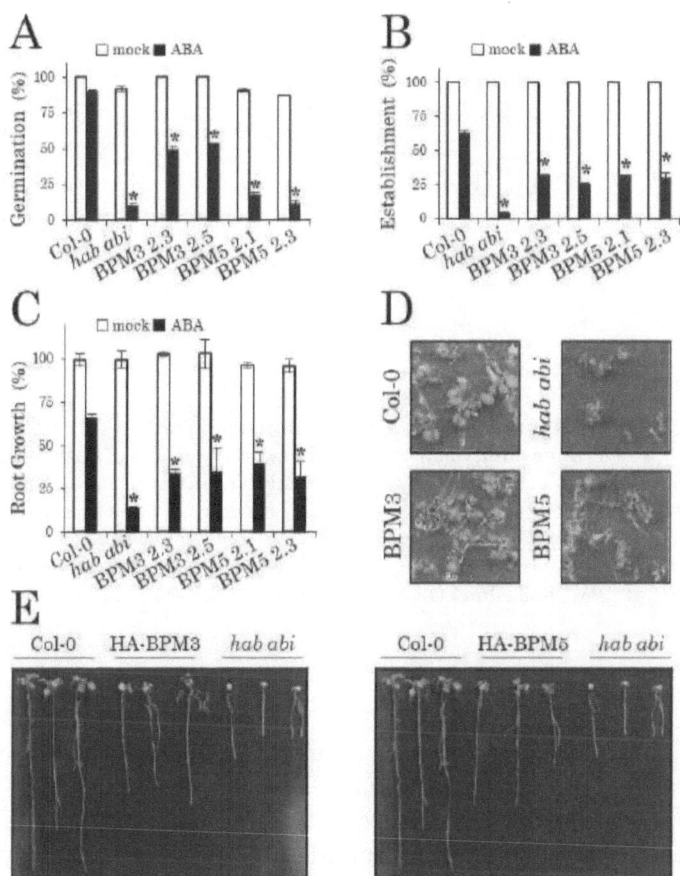

Figure 38. Overexpression of BPM3 and BPM5 leads to enhanced ABA sensitivity compared to the Col-0 wild type. (A) Enhanced sensitivity to ABA-mediated inhibition of seed germination in BPM3 and BPM5 OE lines compared to Col-0. Data of the ABA-hypersensitive mutant line *hab1-1 abi1-2* was also recovered. Approximately 100 seeds of each genotype (two independent experiments) were sown on MS plates lacking or supplemented with 0.5µM ABA. Germination (radicle emergence) was scored after 3d. Data are means ± SD. Asterisks indicate p < 0.05 (Student's t test) when comparing data with Col-0 in the same conditions. (B) Enhanced sensitivity to ABA-mediated inhibition of seedling establishment. The same seedlings of (A) were scored after 8d for the presence of both, green cotyledons and the first pair of true leaves. (C) Enhanced sensitivity to ABA-mediated inhibition of root growth. 20 seedlings of each genotype (three independent experiments) were growth on vertical MS plates for 4d. Then, they were transferred to vertical MS plates lacking or supplemented with 10µM ABA. Root growth was scored after 10d. (D) Pictures taken of representative seedlings of (B) with 0.5µM ABA. (E) Pictures taken of representative seedlings of (C) with10µM ABA.

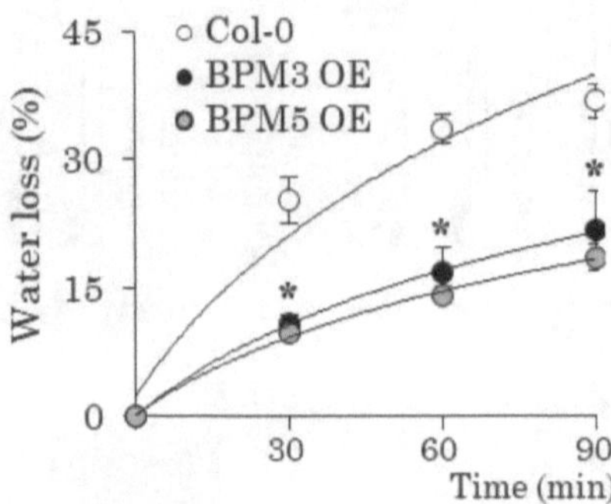

Figure 39. Diminished water loss in detached leaves of BPM3 and BPM5 OE lines compared to the Col-0 wild type. Loss of fresh weight was measured at different times in 15d old leaves submitted to the drying atmosphere of a laminar flow hood. Data are means of three independent experiments ±SD (n=10 per experiment). The lines represent exponential trendlines. Asterisks indicate p < 0.05 (Student's t test) when comparing data of BPM3 and BPM5 OE lines with Col-0 at the same time points.

Finally, we performed a drought experiment to further analyse the water loss phenotype. We observed that BPM5 OE plants showed enhanced drought resistance under greenhouse conditions compared to the Col-0 wild type (**Figure 40**). Once again we added the *hab1-1 abi1-2* mutant line as an example, in this case of a line with a high drought resistance. More than half of the BPM5 OE plants tested survived. We also performed the experiment with BPM3 OE plants but they didn't present a higher drought resistance. This can have to do with the difference of BPM protein present in each line (**Figure 32**).

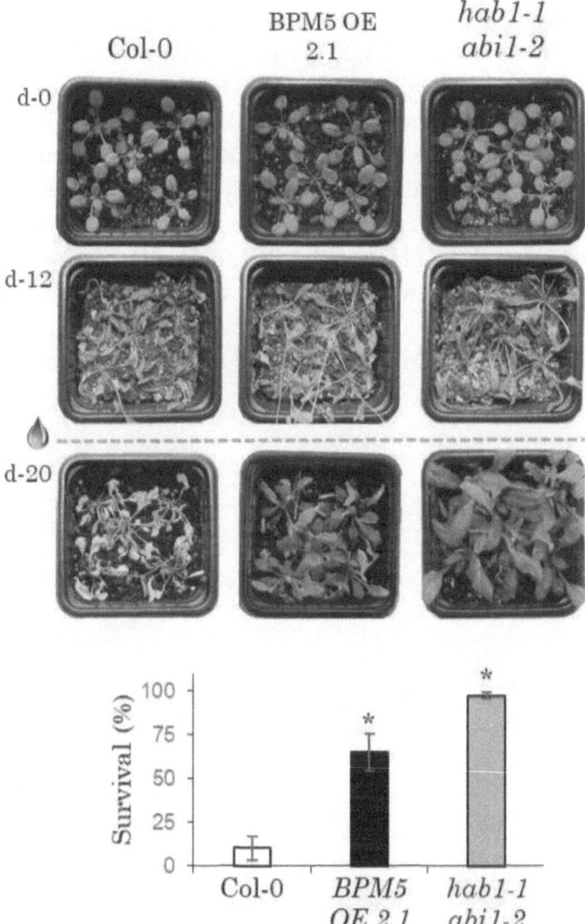

Figure 40. Enhanced drought resistance of BPM5 OE plants compared to wild type Col-0. We stopped the irrigation of 2w old plants (d0) for 12d. Then, we recover the irrigation (d12). After 8d, the survival was scored (d20). Data are means of three independent experiments ± SD (n=10 per experiment). Asterisks indicate $p < 0.05$ (Student's t test) when comparing BPM5 OE line 2.1 and *hab1-1 abi1-2* with Col-0 at the same time points.

Finally, we wanted to know more about the differences on BPM3 and BPM5 OE lines expression and the possible effect of ABA and MG-132 on them. For this, we submitted two of each BPM3 and BPM5 OE lines to ABA, MG and ABA+MG treatments and analysed the BPMs levels by immunoblot analysis. We observed a slightly increase in BPM3 accumulation in response to ABA and a higher accumulation in presence of MG-132 (**Figure 41**). These results suggest that BPM3 is being degraded by the 26S proteasome and is stabilized by ABA. However, BPM5 levels did not change among treatments, which imply that BPM5 is not degraded by the proteasome and thus, explains the higher levels of accumulation of BPM5 over BPM3 OE lines (**Figure 32**).

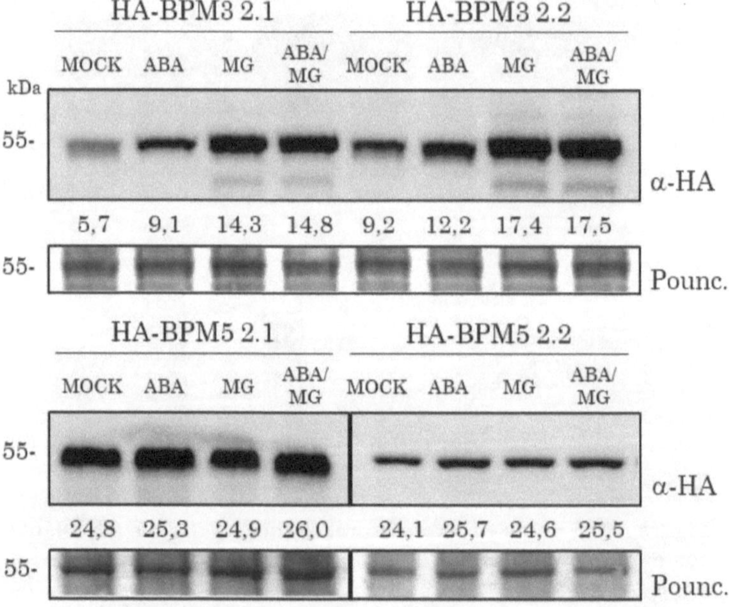

Figure 41. ABA and MG-132 stabilize BPM3. Immunoblot analysis of HA-tagged BPM3/5 proteins levels in *A. thaliana* BPM3/5 OE lines after mock, 50μM ABA, 50μM MG-132 or ABA + MG-132 3h treatments. Numbers represent percentage of total signal recover from the analysis. Ponceau staining of Rubisco was used as a protein loading control.

bpm3 bpm5 loss-of-function shows reduced sensitivity to ABA.

BPMs belong to a family of six members in *A. thaliana*. Which are required for normal plant development (Lechner et al., 2011). This was revealed by the phenotypic analysis of an *amiR-bpm* mutant line impaired in the transcript expression of *BPM1, 4, 5* and *6*. Leaf shape, leaf size and stem elongation were affected in this line. As well as the overall stature of the plants. Also was observed a severe phenotype in flower development and reduced pollen viability.

Despite this phenotype, we could confirm that detached leaves of *amiR-bpm* plants showed higher water loss than Col-0 (**Figure 42**). These plants lost ~25% more water after 4h than Col-0.

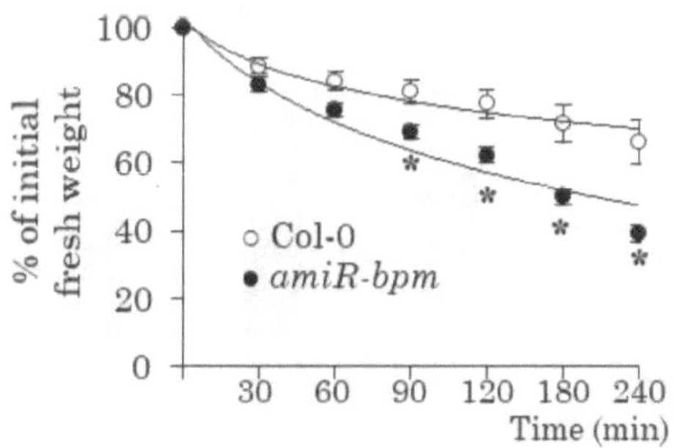

Figure 42. Enhanced water loss in detached leaves of *amiR-bpm* mutant line compared to the Col-0 wild type. Loss of fresh weight was measured at different times in 15d old leaves under laboratory conditions (~25°C, ~40% air relative humidity). Data are means of three independent experiments ±SD (n=10 per experiment). The lines represent exponential trendlines. Asterisks indicate $p < 0.05$ (Student's t test) when comparing data of the *amiR-bpm* line with Col-0 at the same time points.

Additionally, the *amiR-bpm* line also displayed lower leaf temperature than Col-0 (**Figure 43**). The leaves of *amiR-bpm* plants were around 1.5°C colder. These data altogether suggest that ABA-induced stomatal closure might be impaired in this line.

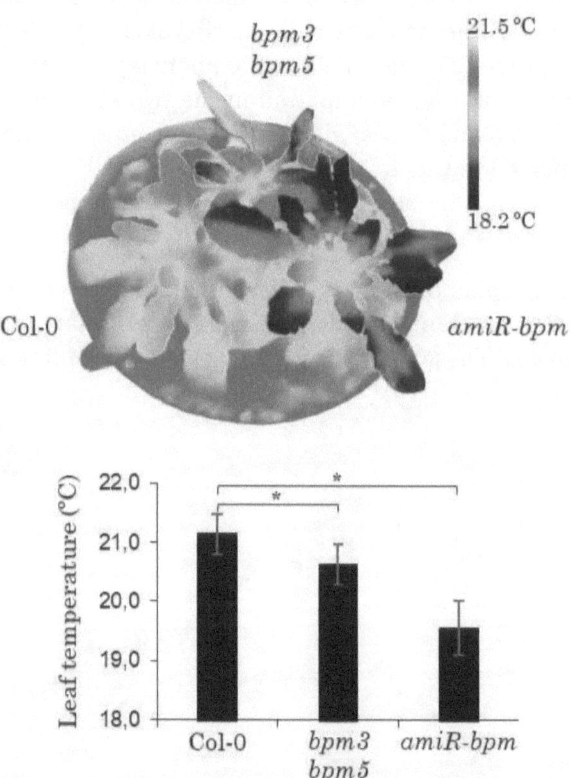

Figure 43. *bpm3 bpm5* and *amiR-bpm* mutant lines are colder than the wild type Col-0. False colours infrared images of Col-0, bpm3 bpm5 and amiR-bpm plants representing leaf temperature. Temperature was quantified by infrared thermal imaging. Data are means of three independent experiments ± SD (n=5 plants per experiment; n=15 measured points per plant). Asterisks indicate p < 0.05 (Student's t test) when comparing data of the *bpm3 bpm5* and *amiR-bpm* lines with Col-0.

To further investigate this stomatal closure phenotype, stomatal conductance experiments were performed by Dr. Ebe Merilo (**Figure 44**). Well-watered *amiR-bpm* and Col-0 plants have similar pre-ABA stomatal conductance (G$_{ST}$). Both lines showed ABA-induced stomatal closure (**Figure 44A**). However, after ABA treatment, the G$_{ST}$ of the *amiR-bpm* plants was significantly higher than the G$_{ST}$ of Col-0. This indicates that ABA-induced reduction of G$_{ST}$ was impaired in the *amiR-bpm* plants. Then, we grew plants under water deficit and once again we found the same results (**Figure 44B**). No difference in pre-ABA G$_{ST}$ and impaired ABA-response of the *amiR-bpm* line.

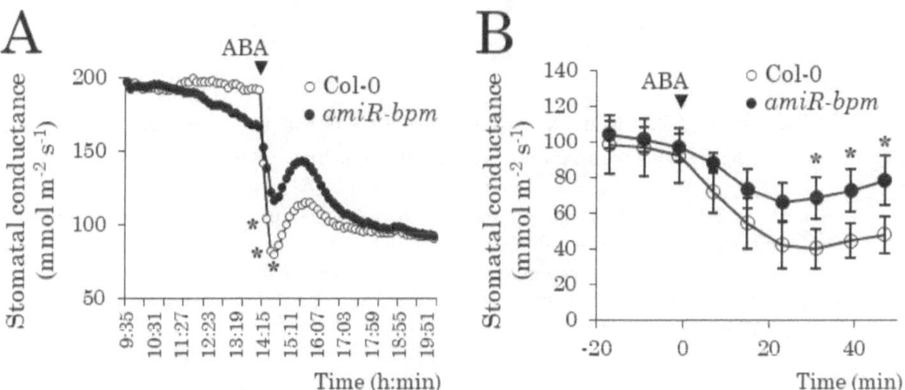

Figure 44. ABA-induced reduction of G$_{ST}$ **is impaired in the** *amiR-bpm line compared to* **the Col-0 wild type.** (A) Course of G$_{ST}$ of well-watered plants along almost half a day. 10μM ABA treatment was performed at 14.00 PM and G$_{ST}$ followed for the next 6h. Data are means (n=5 for Col-0 and n=6 for *amiR-bpm*). SE is less than 10% and is not represented for the sake of clarity. Asterisks indicate p < 0.1 (Student's t test) when comparing data of the *amiR-bpm* line with Col-0 at the same time points. (B) The same experiment as (A) but with plants submitted to drought. The course is reduced to only the prior 20 minutes (min) to the 10μM ABA treatment and the next 50min (n=4).

ABA-induced stomatal closure was not significantly affected in the *bpm3 bpm5* mutant line. This suggest that there is certain functional redundancy of the BPM family in the control of the stomatal aperture (**Figure 45**). As state before, the *amiR-bpm* line presented an impaired ABA-induced reduction of G_{ST} compared to Col-0.

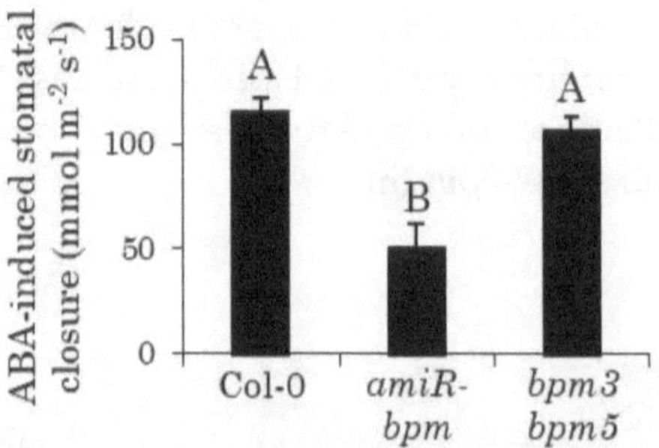

Figure 45. ABA-induced reduction of G_{ST} is not impaired in the *bpm3 bpm5* line compared to the *amiR-bpm* line. Course of G_{ST} of *amiR-bpm* (n=5), *bpm3 bpm5* (n=5) and Col-0 (n=4) was performed as state in Figure 26A. With this data, ABA-induced stomatal closure was calculated as pre-treatment G_{ST} minus G_{ST} after 28min of 10µM ABA treatment. The letters represent different groups formed after performing a One-way ANOVA analysis with post-hoc Tukey test of the data, with a significance level of 0.1.

Given the growth and developmental defects of the *amiR-bpm* line (Lechner et al., 2011), we conducted further phenotypic analysis with the *bpm3 bpm5* double mutant. This line was previously described (**Figure 34**). As described above these plants accumulated more PP2Cs in the roots than Col-0 (**Figure 35**). PP2CA is a PP2C that strongly blocks ABA signaling during germination and early growth (Kuhn et al., 2006; Yoshida et al., 2006). As a result, the *pp2ca-1* mutant line shows the strongest ABA hypersensitivity in seed germination assays compared to the others loss-of-function *pp2c* mutants (Rubio et al., 2009). For this reason, we generated a *pp2ca-1 bpm3 bpm5* triple mutant for epistatic analysis. Then, we performed a seedling establishment assay, in which *bpm3 bpm5* showed reduced ABA sensitivity (**Figure 46**). Around ~90% of the seedlings of *bpm3 bpm5* established after 9d in presence of 1μM ABA, whereas only ~20% of wild type seedlings established. Neither *pp2ca-1* mutant line nor the *pp2ca-1 bpm3 bpm5* triple mutant line established in the same conditions. Therefore, *pp2ca-1* was epistatic to *bpm3 bpm5*.

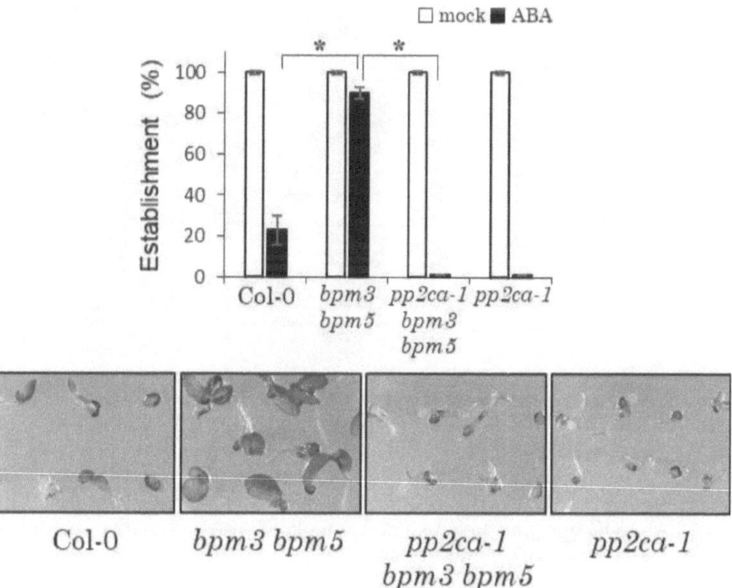

Figure 46. Diminished sensitivity to ABA-mediated inhibition of seedling establishment in *bpm3 bpm5* line compared to Col-0. Data of the ABA-hypersensitive mutant line *pp2ca-1* was also recovered. Also, the *pp2ca-1* mutation abolishes the ABA-insensitive phenotype in the triple mutant *pp2ca-1 bpm3 bpm5*. Approximately 100 seeds of each genotype (two independent experiments) were sown on MS plates lacking or supplemented with 1μM ABA. Seedling establishment was scored after 9d for the presence of both, green cotyledons and the first pair of true leaves. Data are means ± SD. Asterisks indicate p < 0.05 (Student's t test) when comparing data of *bpm3 bpm5* with Col-0 and *pp2ca-1 bpm3 bpm5* in the same conditions. Pictures show representative seedlings of each genotype.

Moreover, we performed a root growth assay. The *bpm3 bpm5* mutant line showed again reduced ABA sensitivity (**Figure 47**). We added the 35S:HAB1 OE line as an example of ABA-insensitive seedlings. The roots of *bpm3 bpm5* seedlings growth around ~15% more than the ones of Col-0.

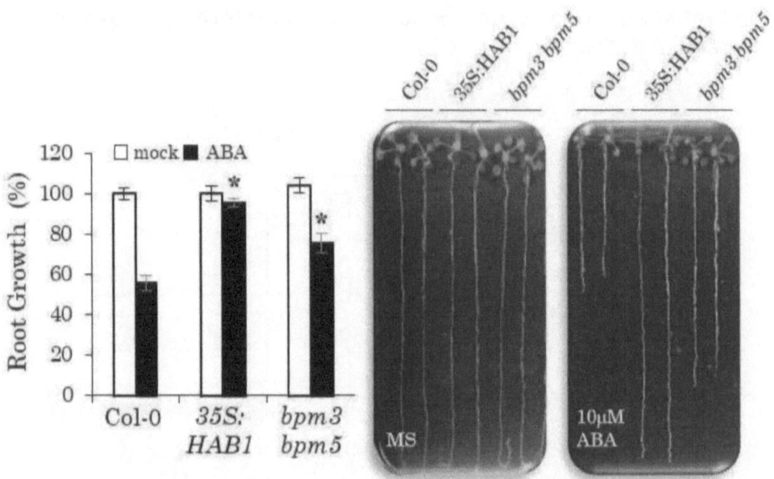

Figure 47. Diminished sensitivity to ABA-mediated inhibition of root growth in *bpm3 bpm5* line compared to Col-0. 20 seedlings of each genotype (two independent experiments) were growth on vertical MS plates for 4d. Then, they were transferred to vertical MS plates lacking or supplemented with 10μM ABA. Root growth was scored after 10d. The 35S:HAB1 OE line was added as an example of ABA-insensitive seedlings. Data are means ± SD. Asterisks indicate p < 0.05 (Student's t test) when comparing data of *bpm3 bpm5* or 35S:HAB1 OE with Col-0 in the same conditions. Pictures show representative seedlings of each genotype.

Transpiration was also monitored using infrared thermography. *bpm3 bpm5* plants showed lower leaf temperature than Col-0 (**Figure 43**). But they were not as fresh as the *amiR-bpm* plants. *bpm3 bpm5* plants were around 0.5°C colder than Col-0.

Then, we analysed the expression of ABA-responsive genes. We did it by RT-PCR at endogenous ABA levels in *bpm3 bpm5* and Col-0. *RAB18*, *RD29B*, *KIN10* and *RD22* were less expressed in the *bpm3 bpm5* mutant line (**Figure 48**). They were around ~50% less expressed than in Col-0.

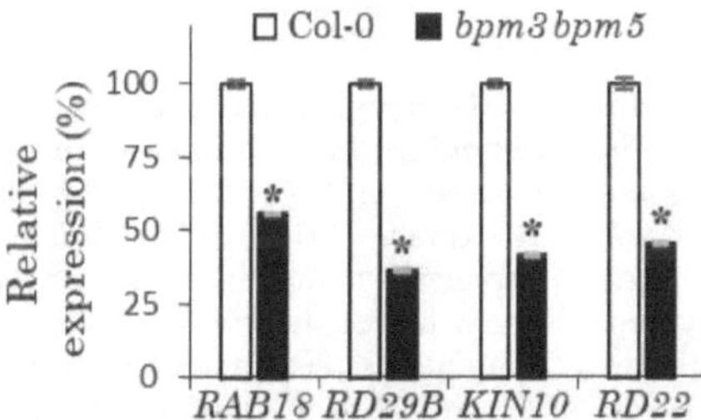

Figure 48. Reduced expression of the ABA-responsive genes *RAB18*, *RD29B*, *KIN10* and *RD22* in *bpm3 bpm5* mutant line compared to Col-0. RT-qPCR analysis was performed from mRNAs obtained from 2w old seedlings. Data are means of three independent experiments ± SD. Asterisks indicate p < 0.05 (Student's t test) when comparing data of *bpm3 bpm5* with Col-0.

Lastly, we also analysed the induction of the ABA-responsive promotor *RD29B* after ABA treatment in transfected protoplasts of *bpm3 bpm5* and Col-0. We measured ABA-induced *LUC* expression driven by *pRD29B* and found a reduced expression in the *bpm3 bpm5* mutant line compared to Col-0 (**Figure 49**). We also transfected with *GUS* for normalization, with the ABA-responsive transcription factor *ABF2* and with or without the *OST1* kinase. ABF2 and OST1 served to promote the ABA pathway, so we could collect more clear information. In both cases we found a reduce expression of *LUC* in the *bpm3 bpm5* protoplasts compared to Col-0. Around ~50% when only the *ABF2* effector plasmid was transfected. With the help of *OST1* the difference was even higher, ~60%.

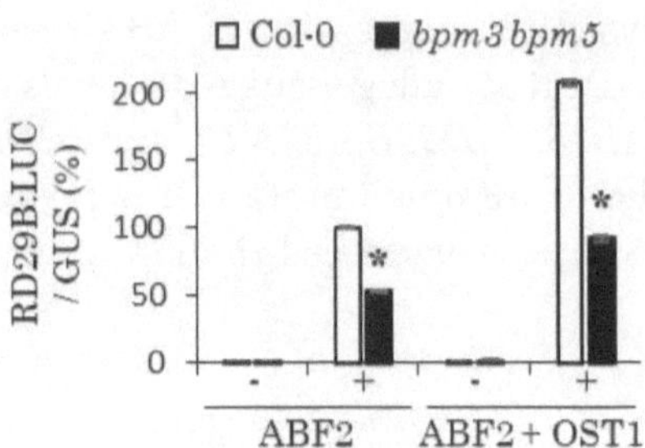

Figure 49. ABA-induced LUC expression driven by *RD29B* promotor was diminished in *bpm3 bpm5* mutant compared to Col-0. Firstly, we did protoplast of *bpm3 bpm5* and Col-0. Then, they were transfected with *RD29B:LUC*, that will only express if the ABA pathway is on. *GUS* for normalization. The ABA-responsive transcription factor *ABF2*, to induce the ABA pathway and with or without the kinase *OST1*, that induce even more the ABA pathway. Protoplasts suspensions were incubated for 6h after transfection. 3h after transfection, half of the protoplasts were incubated with 5mM exogenous ABA (represented with a +). LUC signal was measured and normalized against GUS signal. Data are means of three independent experiments ± SD. Asterisks indicate p < 0.05 (Student's t test) when comparing data of *bpm3 bpm5* with Col-0.

Once the interaction of PP2CA with BPM3 and BPM5 was clear and also the resultant degradation of the PP2CA. Our last objective was to investigate if the PP2CA was previously ubiquitinated. For this, we performed transient expression of PP2CA-GFP in *N. benthamiana* leaves to detect the *in vivo* ubiquitination of the PP2CA. When we co-expressed HA-BPM3 and FLAG-Ub, PP2CA-GFP was more ubiquitinated (**Figure 50**). The FLAG-Ub was induced by β-estradiol to

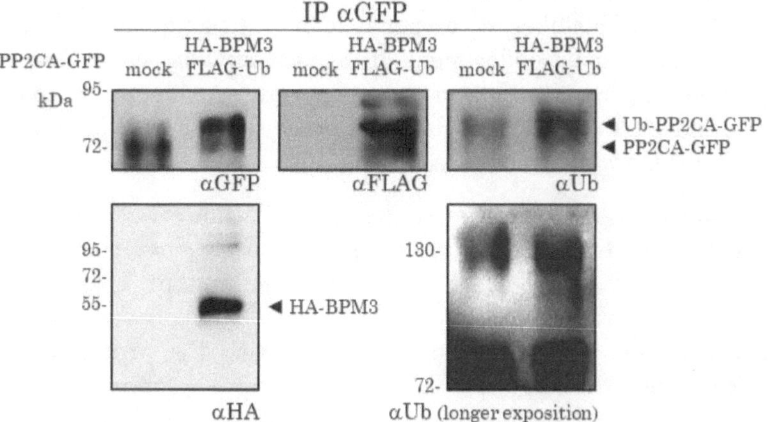

Figure 50. *In vivo* **ubiquitination of PP2CA is enhanced by BPM3 in N. benthamiana.** *A. tumefaciens* encoding PP2CA-GFP was co-infiltrated in leaf cells alone (mock) or co-expressing constitutively HA-BPM3 and FLAG-Ub induced by β-estradiol. 2d after agroinfiltration, 1.5cm disk samples were collected. Then, they were incubated for 16h in 100μM β-estradiol to induce the expression of FLAG-Ub. Protein extracts were immunoprecipitated using α-GFP and analysed by immunoblot with four different antibodies. α-GFP to detect the PP2CA-GFP. α-HA to detect the HA-BPM3. α-FLAG to detect the FLAG-Ub. And α-Ub to detect all forms of ubiquitination. Two pictures of the membrane after incubating with α-Ub are presented. One with a normal exposition and another with a much longer exposition to be able to see the polyubiquitinated forms of the PP2CA-GFP.

begin its expression when required. We performed and IP α-GFP with the protein extracts. Which then were analysed by immunoblot with four different antibodies (α-GFP, α-HA, α-FLAG and α-Ub). Incubating with α-GFP we detected the immunoprecipitated PP2CA-GFP in the mock samples. But when the three proteins were co-expressed and the HA-Ub was induced, PP2CA-GFP mostly shifted to the mono-ubiquitinated form. Incubating with α-HA we saw the co-immunoprecipitated HA-BPM3, that was already described (**Figure 28**). Moreover, incubating with α-FLAG, we could detect the incorporation of FLAG-Ub in the immunoprecipitated PP2CA-GFP. This was confirmed using the α-Ub antibody, to detect all forms of ubiquitinated proteins. In the mock samples, some ubiquitinated forms of PP2CA-GFP are detected incubating with α-Ub (but not with α-FLAG). This likely corresponds to proteins ubiquitinated by the endogenous endowment of *N- benthamiana* cells. Also, upon longer exposition of the membrane, polyubiquitinated forms of PP2CA-GFP could be observed when HA-BPM3 and FLAG-Ub were co-expressed.

Finally, we also investigated the ubiquitination of endogenous PP2CA and HAB1 in *A. thaliana* using three genotypes: wild type Col-0, *bpm3 bpm5* and *amiR-bpm* mutant lines. We prepared protein extracts from the roots of 10d old seedlings. Previously, they were incubated 3h with the proteasome inhibitor MG-132 and ABA. Then, total proteins were incubated with Ub-binding p62 agarose or with agarose lacking p62 as a negative control. The p62 agarose binds all ubiquitinated proteins on the sample. Next, by immunoblot analysis with α-PP2CA or α-HAB1 antibodies, we could detect the endogenous PP2CA or HAB1 and theirs mono-ubiquitinated forms. We already knew that non-ubiquitinated PP2CA and HAB1 accumulated more in the *bpm3 bpm5* mutant line (**Figure 35**). Likewise, we observed it in the *amiR-bpm* line (**Figure 51**, left), using ACTIN for its normalization. So, after normalization, the levels of PP2CA and HAB1 were higher in the mutant lines than in the wild type Col-0. However, after pull-down with the p62 agarose, we recovered more ubiquitinated PP2CA and HAB1 in the Col-0 samples than in the *bpm3 bpm5* and *amiR-bpm* samples (**Figure 51**, central). This suggest that BPMs are required for ubiquitination. As our negative control, we didn't recover ubiquitinated proteins from the agarose lacking the p62 (**Figure 51**, right). Additionally, we incubated the membrane with α-Ub antibody to confirm that the p62 worked as expected, and that we recovered equivalent levels of ubiquitinated proteins. Taken together, these results confirm that the ubiquitination of PP2CA and HAB1 depends on the BPM function.

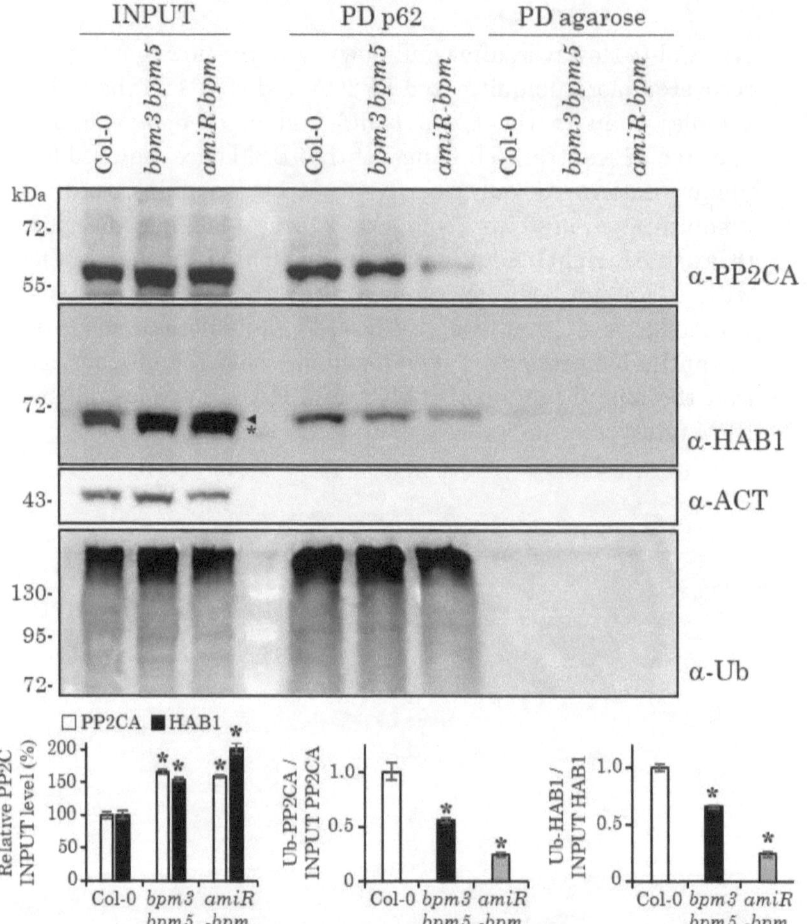

Figure 51. *In vivo* **ubiquitination of PP2CA and HAB1 is diminished in *bpm3 bpm5* and *amiR-bpm* mutants compared to Col-0.** *A. thaliana* seedlings of Col-0, *bpm3 bpm5* and *amiR-bpm* were incubated with 50μM MG-132 and 50μM ABA for 3 h. Then, we did protein extracts of the roots (INPUTs). Which were incubated with Ub-binding p62 agarose to pull-down ubiquitinated proteins, or empty agarose as a negative control. The results were analysed by immunoblot using specific PP2Cs antibodies (α-PP2CA and α-HAB1). This way, we detected non-ubiquitinated and monoubiquitinated forms of both PP2Cs. Actin was analysed as a loading control for the INPUTs with a specific antibody. α-Ub was used to confirm equivalent recovery of ubiquitinated proteins after pull-down with p62 agarose. Left histogram, PP2CA and HAB1 protein INPUT levels in Col-0, *bpm3 bpm5* and *amiR-bpm*. Central histogram, the ratio of Ub-PP2CA / non-ubiquitinated PP2CA obtained respectively from the PD p62 or INPUT samples. Right histogram, the ratio of Ub-HAB1 / non-ubiquitinated HAB1 obtained respectively from the PD p62 or INPUT samples. Data are means of two independent experiments ± SD. Asterisk indicates p < 0.05 (Student's t test) when comparing data of *bpm3 bpm5* and *amiR-bpm* with Col-0 in the same conditions.

RESULTS II: RGLG1

ABA inhibits myristoylation of RGLG1 E3 ligase, inducing its shuttling to promote nuclear degradation of PP2CA.

RGLG1 localizes at the plasma membrane under non-stress conditions.

To begin with, indirect evidence of the myristoylation of RGLG2 was already described (Yin et al., 2007). They obtained this information through *in vitro* transcription-translation experiments performed with radiolabelled myristic acid. Which was incorporated into the wild type protein, but not into the variant with the second glycine replaced by alanine (G2A). In the same work it was reported that RGLG1 is a plasma membrane-associated protein with also a predicted N-terminal myristoylation site. For that reason, we decided to further investigate the subcellular localization of RGLG1.

We generated vectors overexpressing RGLG1-GFP and its RGLG1^{G2A}-GFP variant. Which contains the mutation described above in the predicted N-terminal myristoylation site. Our first approach was transient expression in *N. benthamiana* leaf cells by agroinfiltration. Them we observed that the variant RGLG1^{G2A}-GFP appeared in most of the nucleus of the infiltrated cells (**Figure 52**). While RGLG1-GFP was hardly found in any nucleus. Nuclear localization

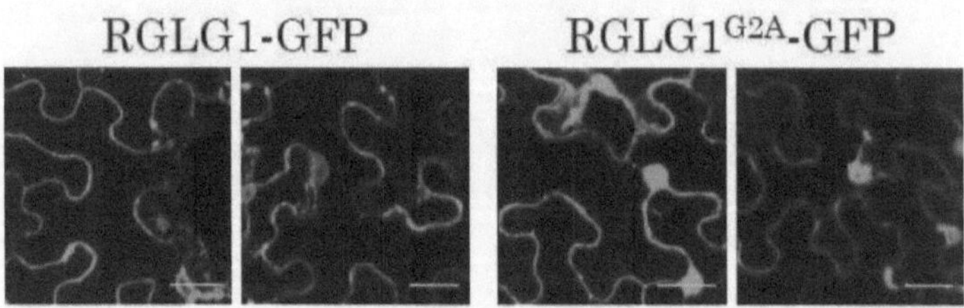

Figure 52. The subcellular localization of RGLG1 depends on the N-terminal myristoylation site. Confocal images showing the subcellular localization of RGLG1-GFP or RGLG1^{G2A}-GFP fusion proteins in transiently transformed N. benthamiana leaf cells. Plants were infiltrated with Agrobacterium tumefaciens harbouring the indicated construct. Scale bars = 30μm.

was confirmed by 4',6-diamidino-2-phenylindole (DAPI) staining (**Figure 53**). Using DAPI staining, we also realized

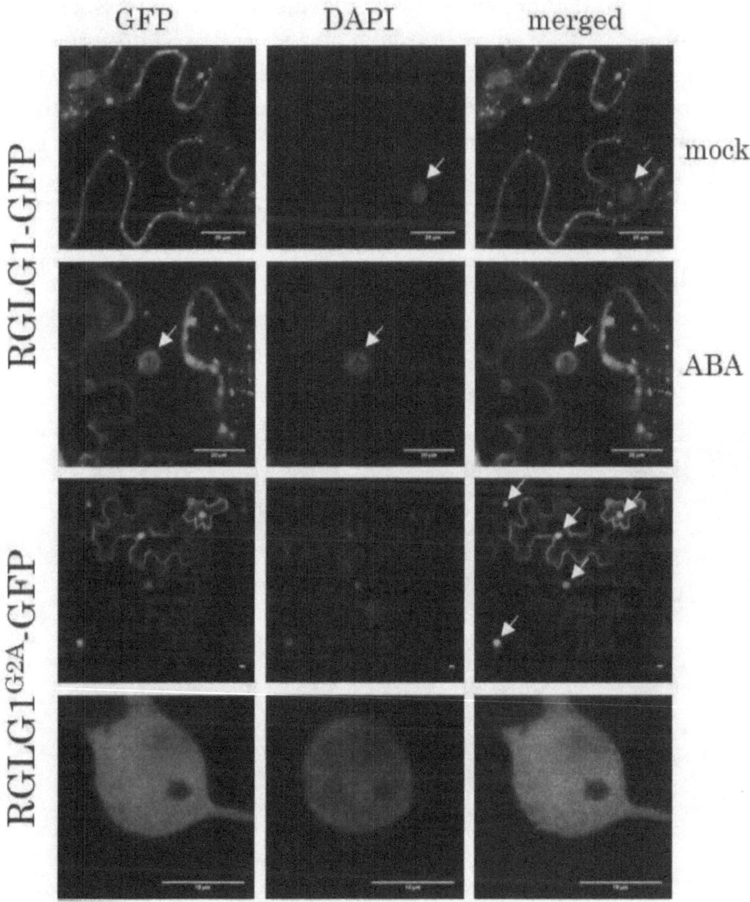

Figure 53. DAPI staining reveals that ABA promotes shuttling of RGLG1-GFP to the cell nucleus, while RGLG1^{G2A}-GFP has a constitutive nuclear localization. ABA treatment was performed 66h after agroinfiltration with constructs encoding RGLG1-GFP or RGLG1^{G2A}-GFP. *N. benthamiana* leaves cells were incubated with mock or 50μM ABA. Then were stained with 5μg/ml DAPI at 72h after agroinfiltration. Approximately 1 mL of the DAPI solution was infiltrated and leaves were analysed immediately. Arrows indicate the cell nucleus. Scale bars = 20μm (general view) or 10μm (zoom).

that after ABA treatment the nuclear staining of RGLG1-GFP was enhanced.

In turgid cells is difficult to distinguish if RGLG1-GFP and RGLG1^{G2A}-GFP are cytosolic or plasma membrane-associated proteins. It is easier in plasmolyzed cells. Therefore, we coinfiltrated each construct with the OFP-TM23 plasma membrane marker and the cells were plasmolyzed with a 500mM NaCl treatment. Colocalization analysis revealed that RGLG1-GFP was localized in the plasma membrane as predicted previously (**Figure 54**). Whereas RGLG1^{G2A}-GFP was not localized in the nucleus but exhibited cytosolic localization.

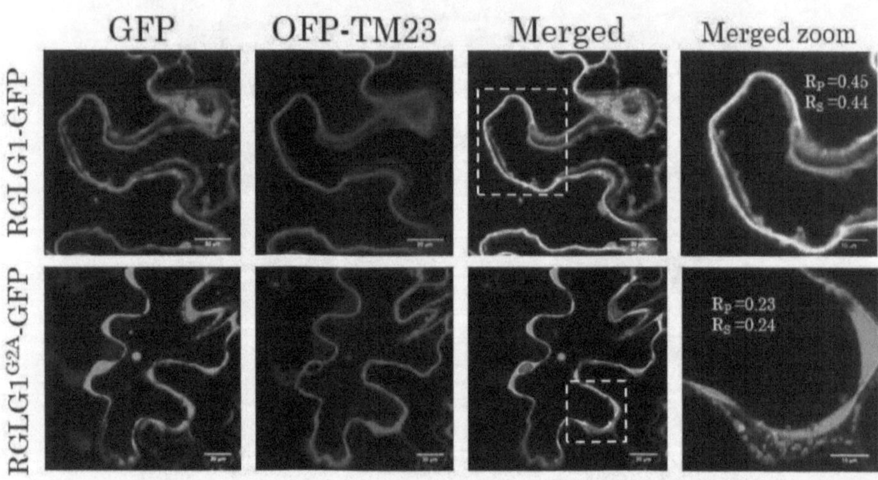

Figure 54. RGLG1-GFP is preferentially localized in the plasma membrane, whereas the RGLG1^{G2A}-GFP variant is found in the cytosol. Confocal images showing the coexpression of RGLG1-GFP or RGLG1^{G2A}-GFP fusion proteins and the plasma membrane marker OFP-TM23 in transiently transformed *N. benthamiana* leaf cells. Prior to the imaging the cells were plasmolyzed with 500µM NaCl. The degree of colocalization between the two fluorescent signals was analysed measuring the values of Pearson's (R$_P$) and Spearman's (R$_S$) correlation coefficients (French et al., 2008). Scale bars = 20/30µm (general view) or 10µm (zoom).

Expression of the corresponding fusion proteins was verified by immunoblot analysis using α-GFP antibodies (**Figure 55**).

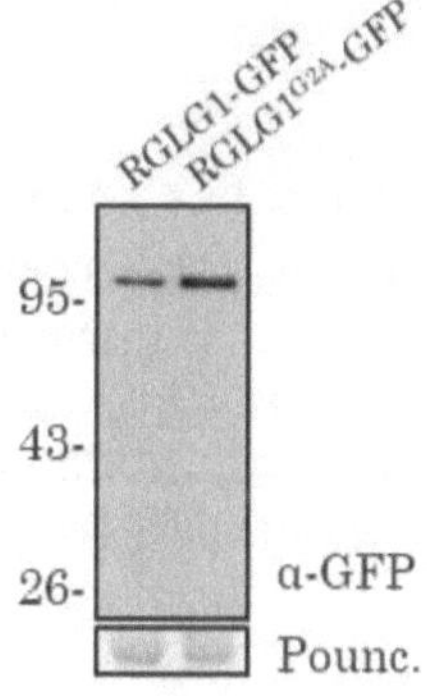

Figure 55. Expression of RGLG1-GFP and RGLG1^{G2A}-GFP in N. benthamiana. Immunoblot analysis confirm the expression of GFP-tagged RGLG1 and RGLG1^{G2A} proteins in *N. benthamiana* transiently transformed plants. Ponceau staining of Rubisco was used as a protein loading control.

RGLG1 is myristoylated under non-stress conditions.

Direct *in vivo* demonstration of myristoylation has been achieved only in a few proteins, and only recent large-scale proteomic approaches have expanded the confirmed myristoylome (Majeran et al., 2018). So, to study the myristoylation of RGLG1, we generated *A. thaliana* transgenic lines that express either RGLG1-GFP or RGLG1[G2A]-GFP (**Figure 56**).

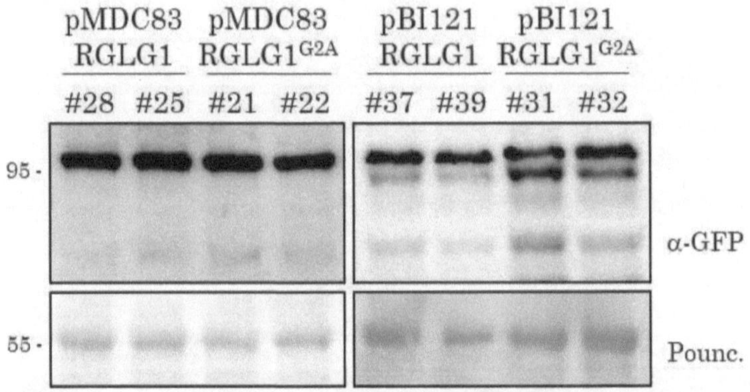

Figure 56. Expression of RGLG1-GFP and RGLG1[G2A]-GFP in *A. thaliana*. Immunoblot analysis confirm the expression of GFP-tagged RGLG1 and RGLG1[G2A] proteins in *A. thaliana* lines transformed with two different vectors, pMDC83 or pBI121. Two independent lines per vector. Ponceau staining of Rubisco was used as a protein loading control.

We already observed that RGLG1-GFP is a plasma membrane-associated protein, whereas its variant RGLG1[G2A]-GFP it's not. So we analysed if this was due to a difference in the myristoylation of both proteins. We immunoprecipitated (α-GFP) samples from both *A. thaliana* transgenic lines. Next, the immunoprecipitates were analyse by immunoblotting with an α-myristic acid antibody. As a result, we detected in vivo myristoylated RGLG1-GFP, whereas lack of myristoylation was found in RGLG1^{G2A}-GFP variant (**Figure 57A**).

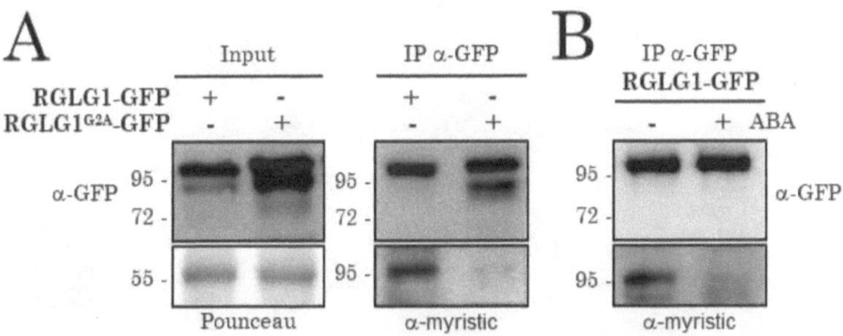

Figure 57. ABA inhibits myristoylation of RGLG1. (A) In vivo myristoylation of RGLG1-GFP. RGLG1-GFP or RGLG1^{G2A}-GFP were expressed in *A. thaliana* transgenic lines. Protein extracts were immunoprecipitated using α-GFP. Then, they were detected by immunoblot analysis using α-GFP or α-myristic acid antibodies. (B) ABA treatment (50μM for 6 h) reduces myristoylation of RGLG1. Analysis of RGLG1-GFP myristoylation was performed as described in (A) with mock- or ABA-treated samples.

Interestingly, ABA treatment dramatically reduced the myristoylation of RGLG1-GFP (**Figure 57B**). This can favour the shuttling of the protein to the nucleus. Indeed, it is known that myristoylated proteins can dynamically relocalize in response to specific signals (Majeran et al., 2018; Turnbull & Hemsley, 2017). The, we performed data mining in public databases and found that expression of NMT1 is diminished by ABA treatment and drought stress (**Figure 58**). NMT1 is the central active enzyme that catalyses myristoylation. So this result is in agreement with the diminished myristoylation of RGLG1. The myristoylome is significantly enriched by signaling and regulatory proteins (Turnbull & Hemsley,

2017). So these data suggest that ABA might enhance the mobilization of these proteins out of the plasma membrane.

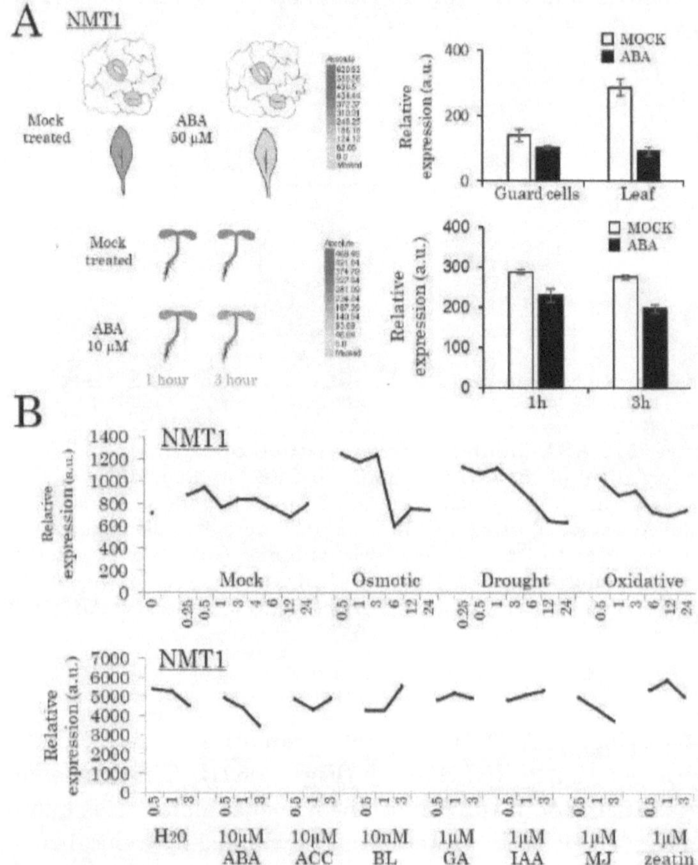

Figure 58. Expression of NMT1 is down-regulated by ABA. (A) Expression of NMT1 in guard cells, leaves and seedlings that were mock- or ABA-treated. Data were visualized using the eFP browser located in http://bar.utoronto.ca/efp_arabidopsis/cgi-bin/efpWeb.cgi (Winter et al., 2007). (B) Relative expression of NMT1 in seedlings that were treated with different hormones (bottom) or submitted (top) to osmotic (300mM mannitol), oxidative (10µM methyl viologen) or drought (dry air stream until 10% loss of fresh weight) stress. Expression data was obtained from Affymetrix microarrays for *A. thaliana* deposited into the AtGenExpress public database (Goda et al., 2008), visualized using the AtGenExpress Visualization Tool located in http://weigelworld.org/resources.html.

RGLG1 shuttles to the nucleus after ABA and salt stress treatment.

Given that ABA diminished myristoylation of RGLG1-GFP, we wanted to know more about its shuttling to the nucleus (**Figure 53**). In this case, we used the *A. thaliana* lines we described above. We examined the subcellular localization of RGLG1-GFP in hypocotyl and root cells, which allows an easy identification of the cell nucleus by bright-field microscopy and where the ABA plays an important role (Belda-Palazon et al., 2018; Wu et al., 2016). In hypocotyl cells we found that RGLG1-GFP was localized in the plasma membrane and in small vesicles found close or associated with the plasma membrane (**Figure 59**). This agrees with our previous experiments in *N. benthamiana* and the role of RGLG1 in K63-linked polyubiquitination (Romero-Barrios & Vert, 2018; Yin et al., 2007). However, after a 50µM ABA treatment for 6h, RGLG1-GFP was also localized in the nucleus (**Figure 59**). We also obtained similar results in root cells (**Figure 60**). To rule out the possibility that DAPI staining might itself affect the localization of RGLG1-GFP, we also identified nucleus by bright-field microscope analysis.

After analysing hypocotyl and root cells, we proceed to the leaves. In this case we could perform a quantification of how many nuclei presented RGLG1-GFP or RGLG1[G2A]-GFP. The number of nuclei decorated with RGLG1-GFP after ABA treatment was 4 fold higher than in mock-treated plants (**Figure 61**). While RGLG1[G2A]-GFP show constitutive nuclear localization. As RGLG1 expression is induced by ABA and abiotic stress (**Figure 7** & **Figure 8**), we analysed whether salt stress affected RGLG1 localization. Additionally, we also performed a calcium treatment because abiotic stress simultaneously increases ABA and calcium concentrations (Edel & Kudla, 2016). Both treatments promoted the shuttling of RGLG1-GFP to the nucleus (**Figure 61**). Neither ABA, salt or calcium treatments affected the cellular distribution of RGLG1^{G2A}-GFP.

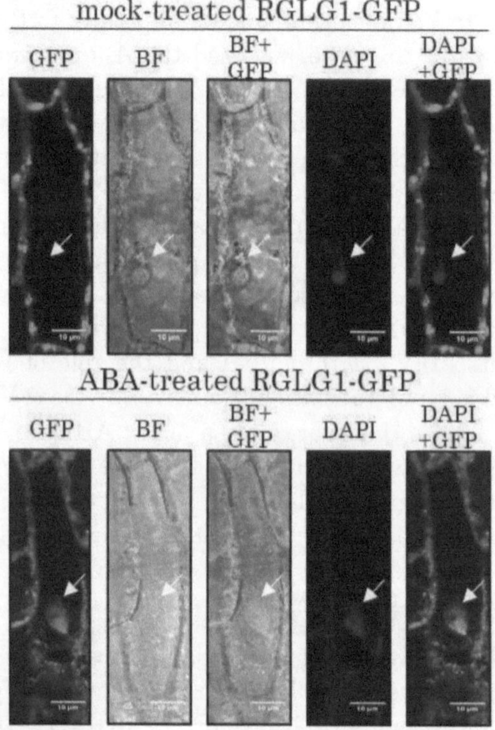

Figure 59. Nuclear localization of RGLG1 in *A. thaliana* hypocotyl cells was promoted by 50µM ABA treatment for 6h. Confocal images of the subcellular localization of RGLG1-GFP in mock- and ABA-treated plants. The GFP channel shows the subcellular localization of RGLG1-GFP. The bright-field (BF) microscope imaging served to identify nuclei. Nuclei were confirmed by DAPI staining. Nuclei are indicated by arrows. Scale bars = 10µm.

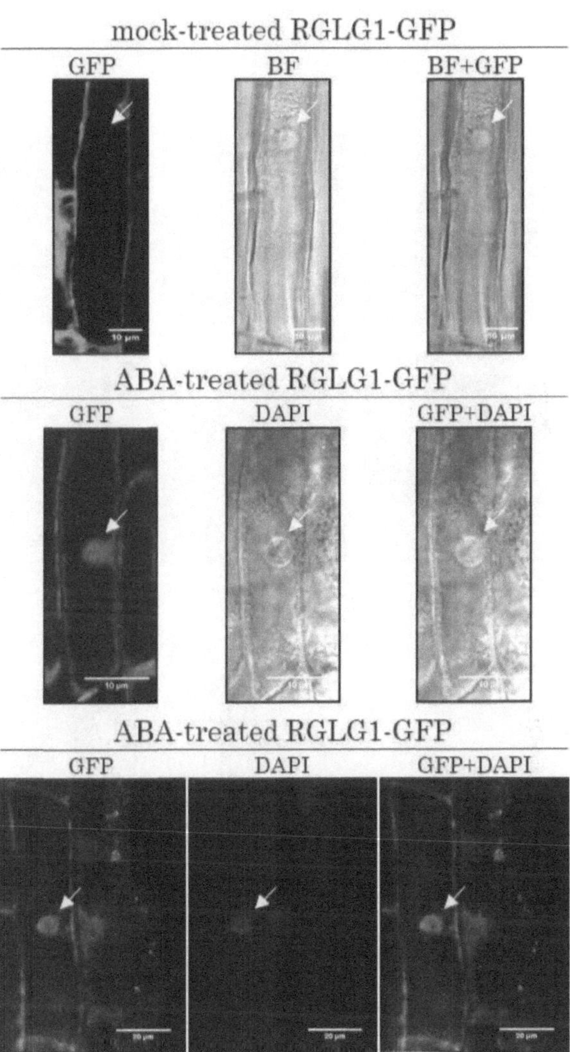

Figure 60. Nuclear localization of RGLG1 in *A. thaliana* root cells was promoted by 50μM ABA treatment for 6h. Confocal images of the subcellular localization of RGLG1-GFP in mock- and ABA-treated plants. The GFP channel shows the subcellular localization of RGLG1-GFP. The bright-field (BF) microscope imaging served to identify nuclei. Nuclei were confirmed by DAPI staining. Nuclei are indicated by arrows. Scale bars = 10/20μm.

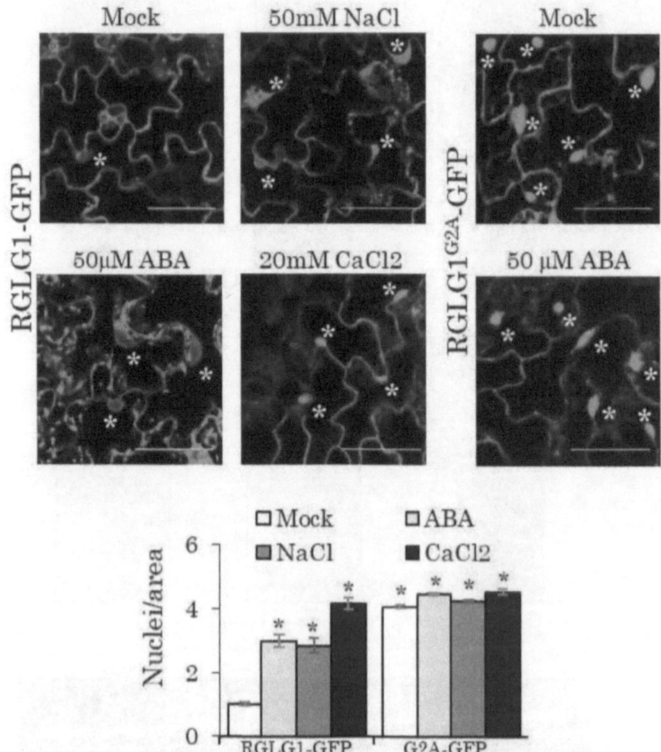

Figure 61. Nuclear localization of RGLG1 in *A. thaliana* leaf cells was promoted by 50μM ABA, 50mM salt and 20mM calcium treatment for 6h. Confocal images of the subcellular localization of RGLG1-GFP and RGLG1^{G2A}-GFP in *A. thaliana* leaf cells. RGLG1^{G2A}-GFP shows constitutive localization in the nucleus. Nuclei are labelled with asterisks. Scale bars = 50mm. The histogram indicates the relative number of nuclei decorated by RGLG1-GFP or RGLG1^{G2A}-GFP per area. Nuclei counting was done in sections of 40.000 mm^2 (n = 20 fields) from 5 independent plants. Analysed through a full z-series of confocal images. Asterisks in the histogram indicate $p < 0.05$ (Student's t test) when comparing treated with untreated plants (Mock).

Demyristoylation enzymes have not been described in plants. So, we wanted to know if the protein localized in the nucleus after ABA treatment was translocated from the plasma membrane or was synthetized *de novo*. For this, we performed the ABA treatment together with CHX, which inhibits protein translation. We found that CHX did not prevent the shuttling (**Figure 62**).

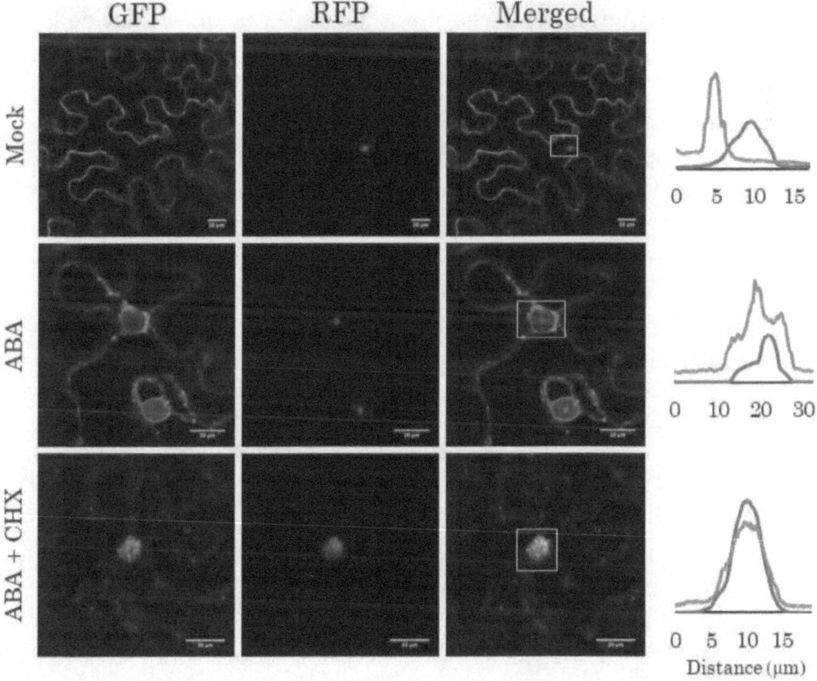

Figure 62. CHX treatment does not prevent ABA-induced shuttling of RGLG1 to the cell nucleus in *N. benthamiana* leaf cells. Confocal images of the subcellular localization of RGLG1-GFP expressed in epidermal leaf cells of *N. benthamiana* that were mock-, 50µM ABA- or 50µMABA + 100µM CHX-treated for 6 h. De novo biosynthesis of RGLG1-GFP was inhibited by the CHX treatment. The GFP channel shows the subcellular localization of RGLG1-GFP. The RFP channel shows the localization of the nucleolar marker Fib-RFP. CHX treatment led to simultaneous staining of nucleolus and nucleoplasm by Fib-RFP. Scale bars = 20µm. The intensity profiles of GFP (green) and RFP (red) fluorescence were measured along the indicated distance (µm) of the selected yellow boxes.

Therefore, *de novo* synthesis of RGLG1 is not required and translocation from the plasma membrane occurs in presence of ABA. Interestingly, we could observe a reduction of fluorescent signal in the plasma membrane after CHX treatment. These results suggest a proteolytic processing of N-myristoylated RGLG1 (Burnaevskiy et al., 2013). Alternative, other mechanisms can also deliver surface proteins to the nucleus (Chaumet et al., 2015). DAPI staining is very aggressive to the cell and did not work well together with the ABA and CHX treatment. For this reason, we decided to use another marker: Fibrillarin-RFP (Fib-RFP) (Herranz et al., 2012). Fib is a component of a small nuclear ribonucleoprotein (Hornacek et al., 2017), and accordingly, Fib-RFP specifically marked the nucleoli of *N. benthamiana* leaf cells. We performed a quantification on how many nuclei presented RGLG1-GFP after each treatment (**Figure 63**). There were 2 fold more nuclei decorated with RGLG1-GFP after the CHX+ABA treatment than in the mock-treated plants. We could also observe that CHX treatment disordered the nucleoli (**Figure 62**).

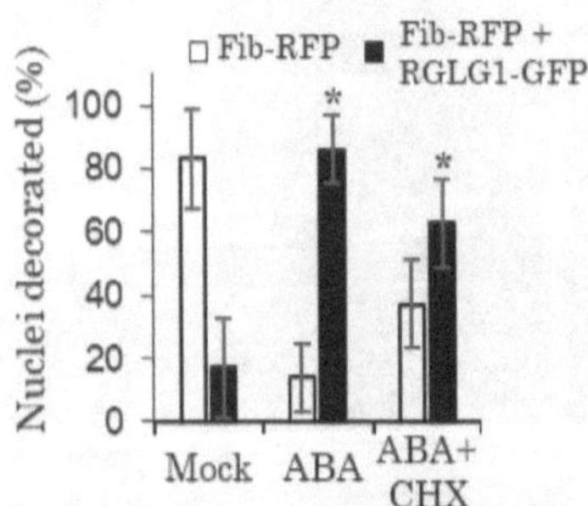

Figure 63. CHX treatment does not prevent ABA-induced shuttling of RGLG1 to the cell nucleus in *N. benthamiana* leaf cells. Percentage of nuclei decorated only by Fib-RFP or simultaneously by Fib-RFP + RGLG1-GFP. Asterisks indicate p < 0.05 (Student's t test) when comparing treated with untreated plants (Mock).

Finally, we were interested in knowing how this affects the sensitivity of the plant to ABA. We compared ABA sensitivity in germination assays of the independent *A. thaliana* lines described above. Constitutive expression of RGLG1[G2A]-GFP, which is strongly localized in the nucleus, produced an enhanced sensitivity to the ABA-mediated inhibition of seed germination compared to RGLG1-GFP (**Figure 64**, left).

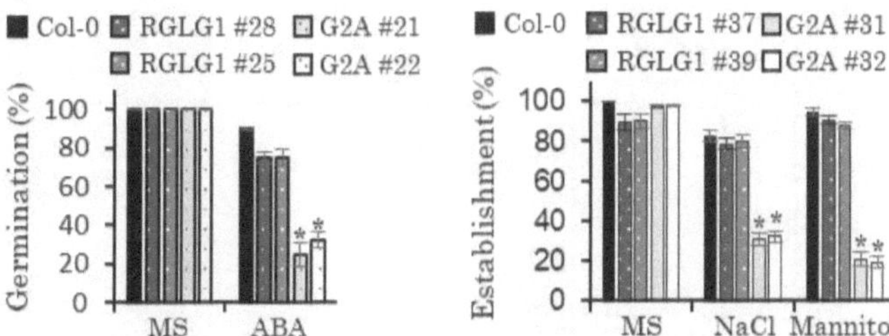

Figure 64. Constitutive expression of RGLG1^{G2A}-GFP leads to enhanced sensitivity to ABA-mediated inhibition of germination (left) or NaCl- and mannitol-mediated inhibition of seedling establishment (right) compared to RGLG1-GFP. Approximately 100 seeds of each genotype (two independent experiments) were sown on MS plates lacking or supplemented with 0.5µM ABA / 50mM NaCl / 200mM mannitol. Germination (radicle emergence) was scored 3d after stratification for wild type Col-0 and two independent lines of pMDC83 RGLG1-GFP and RGLG1^{G2A}-GFP. Seedling establishment was scored 4d after stratification for the presence of both, green cotyledons and the first pair of true leaves. For this we analysed Col-0 and two independent lines of pMDC83 RGLG1-GFP and RGLG1^{G2A}-GFP. Data are means ± SD. Asterisks indicate $p < 0.05$ (Student's t test) when comparing data of RGLG1^{G2A}-GFP with RGLG1-GFP in the same conditions.

And also similar results were obtained for salt and osmotic stress-mediated inhibition of seedling establishment (**Figure 64**, right). Therefore, nuclear localization of RGLG1 is physiologically relevant to mediate the stress response, because presumably facilitates degradation of nuclear PP2Cs (Wu et al., 2016). We also tested an HA-PP2CA OE line, and the double HA-PP2CA RGLG1-GFP OE line. The OE of HA-PP2CA diminished ABA-mediated inhibition of seedling establishment (**Figure 65**). However, the OE of RGLG1-GFP together does abolish this ABA-insensitivity.

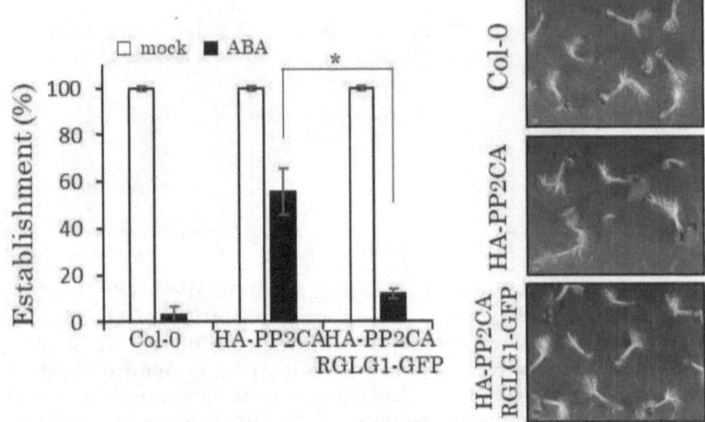

Figure 65. Diminished sensitivity to ABA-mediated inhibition of seedling establishment in PP2CA OE lines is abolished when expressed together with RGLG1. Approximately 100 seeds of Col-0 wild type, HA-PP2CA and HA-PP2CA RGLG1-GFP (two independent experiments) were sown on MS plates lacking or supplemented with 0.5µM ABA. Seedling establishment was scored 4d after stratification for the presence of both, green cotyledons and the first pair of true leaves. Data are means ± SD. The asterisk indicates p < 0.05 (Student's t test) when comparing HA-PP2CA RGLG1-GFP with HA-PP2CA in the same conditions. Pictures show representative seedlings of each genotype under ABA treatment.

Taken together, these results suggest that ABA and salt stress promote the shuttling of RGLG1 to the nucleus. Consequently, PP2CA degradation will be enhanced in response to ABA. Under non-stress conditions the nuclear shuttling of RGLG1 is blocked by myristoylation-mediated attachment to the plasma membrane.

RGLG1 interacts with PP2CA and ABA receptors in the nucleus.

In vitro RGLG1-mediated ubiquitination of PP2CA does not require the presence of ABA or ABA receptors (Wu et al., 2016). However, *in vivo* formation of the receptor-ABA-phosphatase ternary complex might help the recognition of PP2CA by RGLG1. For instance, by generating additional contact points between the E3 ligase and the phosphatase. Firstly, we performed a Y2H assay with ABA receptors and RGLG1 or its variant RGLG1[G2A], to analyse whether could interact or not. PYL4, PYL8 and PYL9 showed interaction with both, RGLG1 and RGLG1[G2A] (**Figure 66**).

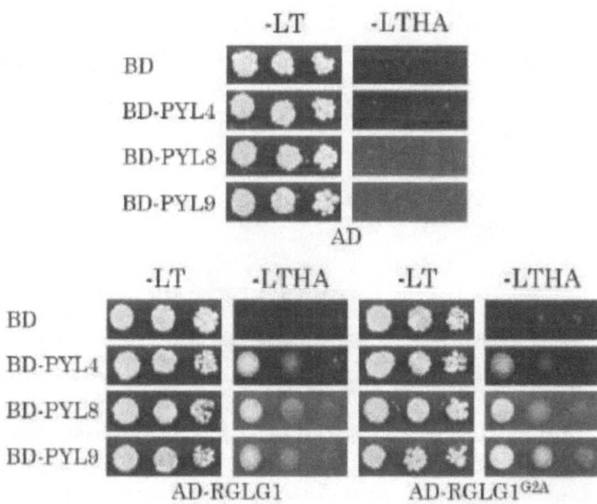

Figure 66. Y2H interaction of RGLG1 or RGLG1[G2A] with PYL4, PYL8 or PYL9. Transformed yeast with BD, BD-PYL4, BD-PYL8 or BD-PYL9 and AD, AD- RGLG1 or AD-RGLG1[G2A], was grown overnight in liquid synthetic (SD) medium lacking Leu and Trp. Dilutions of these cultures were dropped on either control medium lacking Leu and Trp (SD -LT) or selective medium additionally lacking His (SD-LTH). Yeasts were allowed to grow for 3 days (d) at 28°C before interaction was check off.

We focused additional work on the PYL8-RGLG1 interaction. PYL8 is an efficient inhibitor of PP2CA and is predominantly localized in the nucleus (Belda-Palazon et al., 2018). To this end, we generated two double transgenic lines in *A. thaliana*, which coexpress RGLG1-GFP and either HA-PYL8 or HA-PYR1 (as a putative negative control). We tested the interaction between them by Co-IP, and also with the endogenous PP2CA. We found that immunoprecipitated RGLG1-GFP formed complexes with HA-PYL8 and PP2CA in the presence of ABA (**Figure 67**, left). PYL8-ABA-PP2CA ternary complexes were previously described in the presence of ABA (Antoni et al., 2013), so it makes sense to assume that RGLG1 interacts with the whole ternary complex. A lower input of HA-PYL8 and PP2CA is observed in the absence of ABA treatment. This can be explain by the enhanced ubiquitination of PYL8 (Belda-Palazon et al., 2018) and reduced PP2CA transcript levels in the absence of exogenous ABA treatment (Wu et al., 2016). In contrast, HA-PYR1 did not co-immunoprecipitates with RGLG1-GFP (**Figure 67**, right).

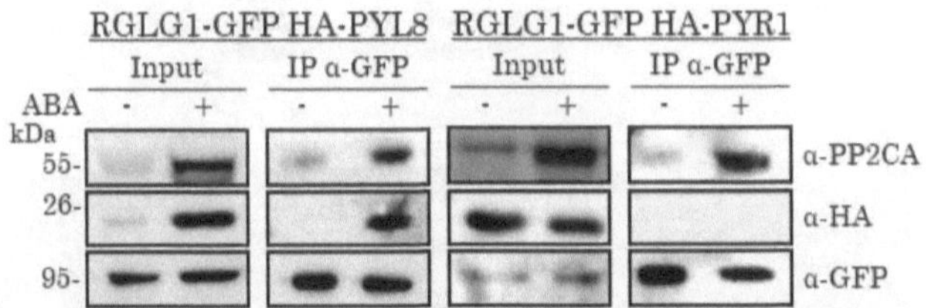

Figure 67. Formation of PYL8-RGLG1-PP2CA ternary complexes in the presence of ABA. CoIP experiments reveal the formation of PYL8-RGLG1-PP2CA complexes after 50µM ABA-treatment in *A. thaliana* double transgenic line co-expressing RGLG1-GFP and HA-PYL8. The IP was done α-GFP. The resultant proteins were analysed by immunoblot using an α-GFP antibody to detect RGLG1 and an α-HA for PYL8 and PYR1. Also, endogenous PP2CA was detected using an α-PP2CA antibody. In contrast to RGLG1 and PYL8, PYR1 does not form such complexes in *A. thaliana* double transgenic lines co-expressing RGLG1-GFP and HA-PYR1.

Once we described the interaction between RGLG1 and PYL4, PYL8 and PYL9, we wandered if this could lead to the ubiquitination of the receptors. To this end, our collaborators Qian Wu and Xu Zhang performed *in vitro* ubiquitination reactions. They added stepwise the E1-E2-E3 ubiquitination cascade. But MBP-RGLG1 (E3) did not ubiquitinated neither GST-PYL4 nor GST-PYL8 (**Figure 68**). Although the cascade was able to ubiquitinate the MBP-RGLG1 (α-FLAG panel).

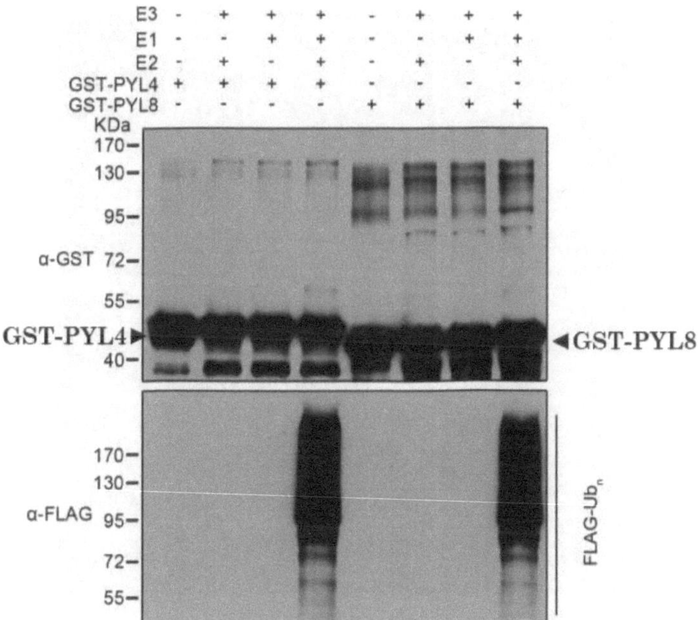

Figure 68. MBP-RGLG1 does not ubiquitinate *in vitro* the PYL4 or PYL8 receptors. A total reaction volume of 30uL was mixed by adding different combinations of MBP-RGLG1 (E3), GST-PYL4 or GST-PYL8, E1 (Sigma-Aldrich), E2 UbcH5b (Enzo Life Sciences) and FLAG-Ub in the ubiquitination buffer. After stopping the reaction, the samples were analysed by immunoblot using an α-GST antibody to detect the receptors and an α-FLAG to detect the ubiquitinated proteins. No ubiquitination of the receptors was observed (α-GST panel). In contrast, *in vitro* autoubiquitination of MBP-RGLG1 (α-FLAG panel) was observed when the E1-E2-E3 components were combined. FLAG-Ub was used to detect the incorporation of Ub into the E1-E2-E3 cascade.

Then, Dr. Borja Belda co-expressed RGLG1-RFP and PP2CA-GFP in *N. benthamiana* leaves by *A. tumefaciens* mediated infiltration. We saw that the higher the amount of PP2CA-GFP that was infiltrated, the more RGLG1-RFP was shuttling to the nucleus, and once there, it did co-localize with the phosphatase (**Figure 69**).

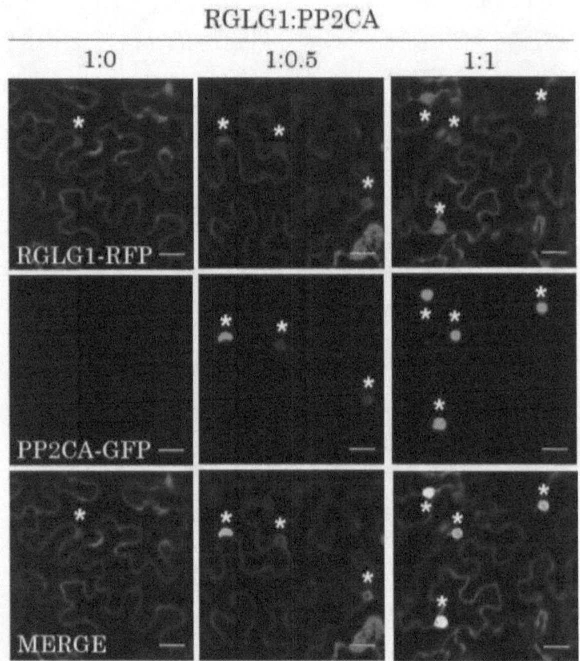

Figure 69. Nuclear localization of RGLG1 is increased after co-expression with PP2CA. Confocal images of the subcellular localization of RGLG1-RFP expressed in epidermal *N. benthamiana* leaf cells in the absence or presence of PP2CA-GFP at different ratios. Increasing amounts of an *A. tumefaciens* that drives the expression of PP2CA-GFP were infiltrated with a constant amount of another agrobacteria driving the expression of RGLG1-RFP. The ratio of the relative concentrations of agrobacteria used in the different coinfiltrations are indicated by numbers (top). Scale bars = 20μm.

It was also performed a quantification of the number of nucleus that expressed RGLG1-RFP in each condition (**Figure 70**). This shuttling to the nucleus might prevent an excessive accumulation of PP2CA after an ABA stimulus, which would facilitate the resetting of the ABA signaling pathway.

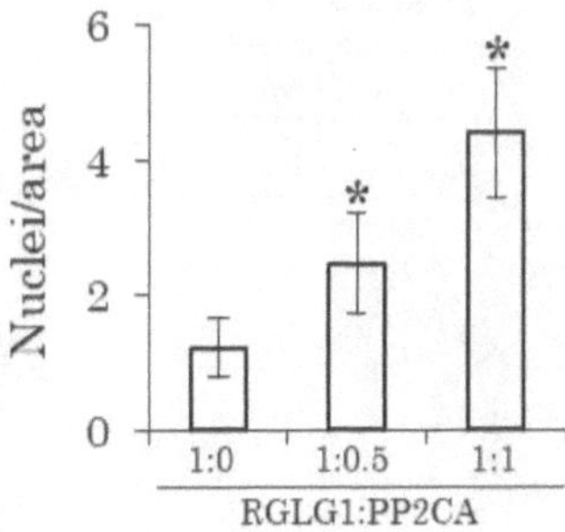

Figure 70. Relative number of nuclei decorated by RGLG1-RFP per area. Nuclei counting was done in sections of 40.000 mm^2 (n = 20 fields) from 5 independent plants. Analysed through a full z-series of confocal images. Asterisks in the histogram indicate $p < 0.05$ (Student's t test) when comparing plants expressing or not PP2CA-GFP.

Finally, we wanted to further investigate the subcellular localization of the interactions between RGLG1 with PYL8 and PP2CA. For this, multicolour BiFC in *N.* benthamiana leaves by *A. tumefaciens* mediated infiltration was performed. SCFP^N-PYL8, RGLG1-SCFP^C and VENUS^N-PP2CA were co-expressed, and PP2CA-RGLG1 interacted once again in the nucleus (**Figure 71**). However, the interaction between PYL8 and RGLG1 only occurs after an ABA treatment. This treatment showed that both complexes, PP2CA-RGLG1 and PYL8-RGLG1, co-localized in the nucleus. So the ternary complex could be formed between the three proteins. Eventually, the empty vectors were tested against all the constructs as negative controls (**Figure 72**).

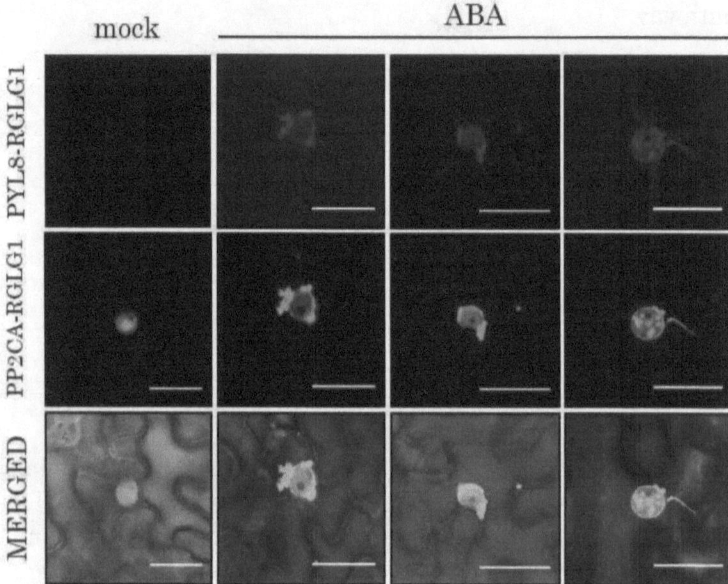

Figure 71. Multicolour BiFC reveals the formation of PYL8-RGLG1-PP2CA complexes in the nucleus of *N. benthamiana* leaves cells after ABA treatment. Confocal images of transiently transformed tobacco epidermal cells co-expressing SCFP^N-PYL8, RGLG1-SCFP^C and VENUS^N-PP2CA. The PYL8-RGLG1 interaction was visualized through reconstitution of the SCFP, whereas the RGLG1-PP2CA interaction gave rise to SCFP^C-VENUS^N fluorescent protein. The interaction of PYL8-RGLG1-PP2CA could be visualized in the nucleus and membrane complexes associated to the nuclear envelope. Constructs were delivered into *N. benthamiana* leaves through *A. tumefaciens* infiltration. Leaves were examined 48-72h after infiltration in the absence of exogenous ABA (mock) or previous a 50µM ABA-treatment for 1 h. Scale bars = 30/40µm for minus or plus exogenous ABA treatment, respectively.

p(MAS):RGLG1-SCYCE
+ pDEST:SCYNE(R)

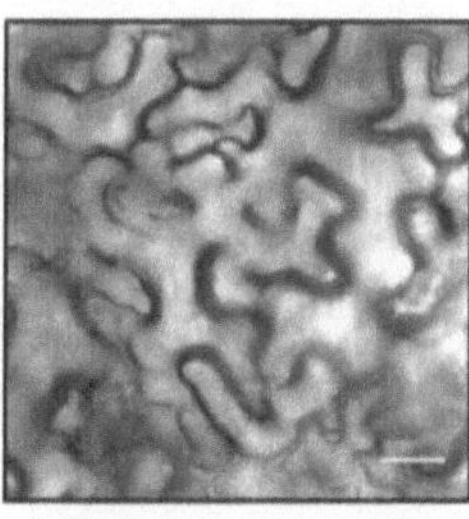

p(MAS):RGLG1-SCYCE
+ pDEST:VYNE(R)

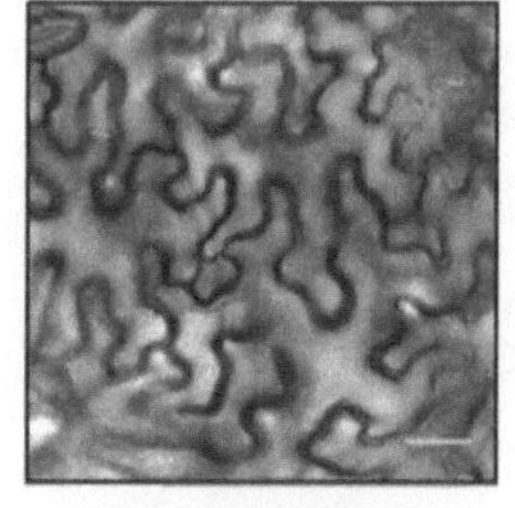

p(MAS):SCYCE +
pDEST:SCYNE(R)-PYL8

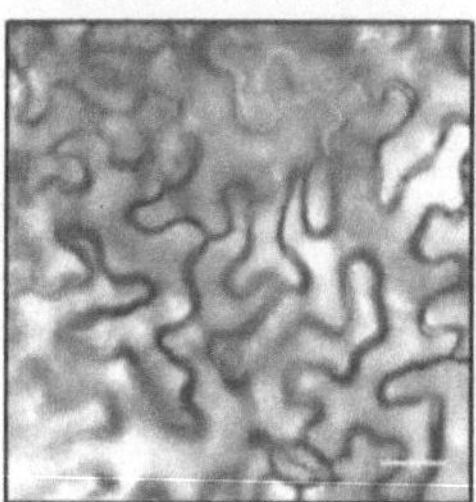

p(MAS):SCYCE +
pDEST:VYNE(R)-PP2CA

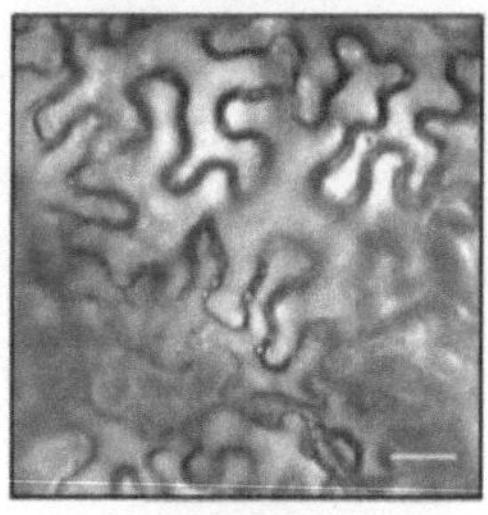

Figure 72. Negative controls of the multicolour BiFC experiment. Different combinations of SCFP^N-PYL8, RGLG1-SCFP^C and VENUS^N-PP2CA were tested against the empty vectors. Constructs were delivered into *N. benthamiana* leaves through *A. tumefaciens* infiltration. Leaves were examined 72h after infiltration. Scale bars = 30µm.

DISCUSSION

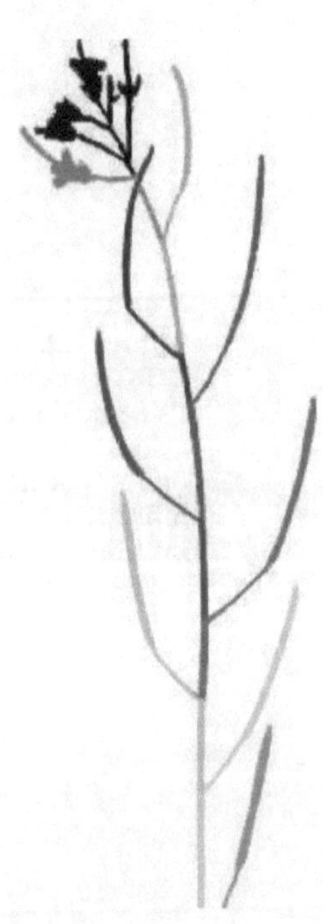

This work combines the study of two types of E3 ligases that contribute to PP2C degradation and ABA signaling. The RING-type RGLG1 E3 ligase was previously reported, but its mechanism of activation in response to ABA has not been elucidated yet. Regarding BPM3 and BPM5 substrate adaptors of CRL3 E3 ligase, this work has discovered the role of CRL3^{BPM} complexes as key regulators of clade A PP2C stability.

The importance of the regulation of PP2C activity for ABA signaling has been well documented. However, other mechanisms, such as the regulation of PP2C protein stability are just emerging. During ABA signaling at least three steps can be distinguished (**Figure 73**): the activation of the ABA response is the first one, followed by the desensitization accomplished by the accumulation of newly synthesized

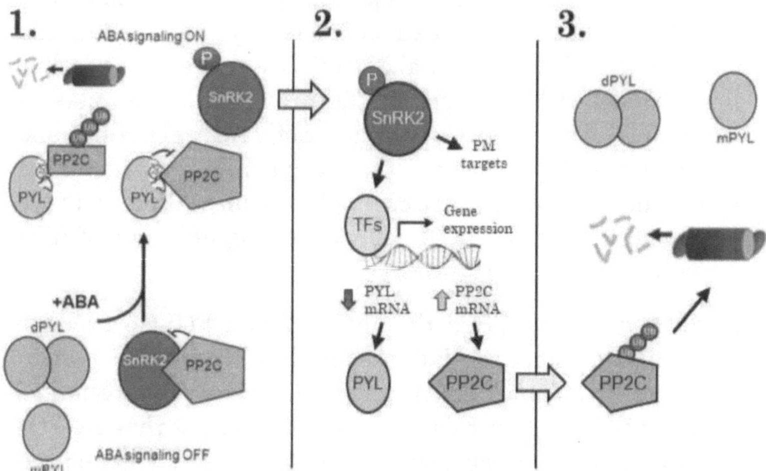

Figure 73. Illustration of the 3 steps of ABA signaling. 1) Activation of the ABA response by formation of the receptor (PYL)-ABA-phosphatase (PP2C) complex releasing the kinase (SnRK2) from PP2C inhibition. SnRK2s are now free to be phosphorylate. PP2Cs are also degraded during this step. 2) Desensitization accomplished by SnRK2 regulation of transcription factors (TFs), which induced accumulation of newly synthesized PP2Cs as a negative feedback mechanism and downregulation of some ABA receptors. 3) Resetting of PP2C proteostasis by activation of E3s that target PP2Cs for degradation.

PP2Cs as a negative feedback mechanism, and finally the resetting of PP2C proteostasis that requires PP2C degradation. All these steps are ABA dependent: formation of the receptor-ABA-phosphatase ternary complexes (Ma et al., 2009; Melcher et al., 2009; Miyazono et al., 2009; Moreno-Alvero et al., 2017; Santiago et al., 2009; Yin et al., 2009), the upregulation of PP2Cs expression (Wu et al., 2016) and downregulation of some ABA receptors (Santiago et al., 2009; Szostkiewicz et al., 2010), and final activation of E3s that target PP2Cs for degradation. The first E3 ligases discovered were PUB12 and PUB13, which required ABA and ABA receptors to promote ABI1 degradation (Kong et al., 2015). Monomeric receptors increased ABI1 ubiquitination by PUB13 in absence of ABA, while ABI1 ubiquitination by PUB13 in presence of PYR1 is strictly dependent on ABA (Kong et al., 2015). The subcellular localization of this interaction has not been reported yet, but it has been postulated that PUB12 can be recruited into the plasma membrane by FLS2 (Lu et al., 2011). Therefore, it is possible that PUB12-ABI1 interact in the proximity of the plasma membrane, where PP2Cs regulate both ABA signaling and different ABA effectors. Alternatively, unidentified membrane linked E3 ligases might regulate PP2Cs in an analogous manner to RSL1-dependent ubiquitination of ABA receptors (Bueso et al., 2014b).

In 2016 it was reported that RGLG1 and RGLG5 promote PP2CA ubiquitination, but the mechanism by which ABA enhances RGLG1/5-PP2CA interaction was not elucidated (Wu et al., 2016). In this work, we discovered that RGLG1 localizes mostly at the plasma membrane under non-stress conditions through a myristoylation modification at the N-terminus (**Figure 52 & Figure 57**). In contrast, RGLG1[G2A] was not myristoylated *in vivo* and it was localized constitutively in both, the nucleus and the cytoplasm. The occasional presence of RGLG1 in the nucleus of some cells was observed; however, after ABA treatment the myristoylation of RGLG1 was reduced by 70% and the number of nuclei with

RGLG1 increased from ~20% to more than 80% (**Figure 59, Figure 60 & Figure 61**). The translocation of RGLG1 to the nucleus was also achieved in presence of calcium and salt stress (**Figure 61**), and was resistant to CHX treatment (**Figure 62 & Figure 63**), indicating that RGLG1 can move from the plasma membrane to the nucleus (**Figure 74**). Salt stress leads to calcium and ABA accumulation (Waadt et al., 2014; Webb et al., 2001), and ABA application has been shown to induce calcium signals (Kim et al., 2010). Furthermore,

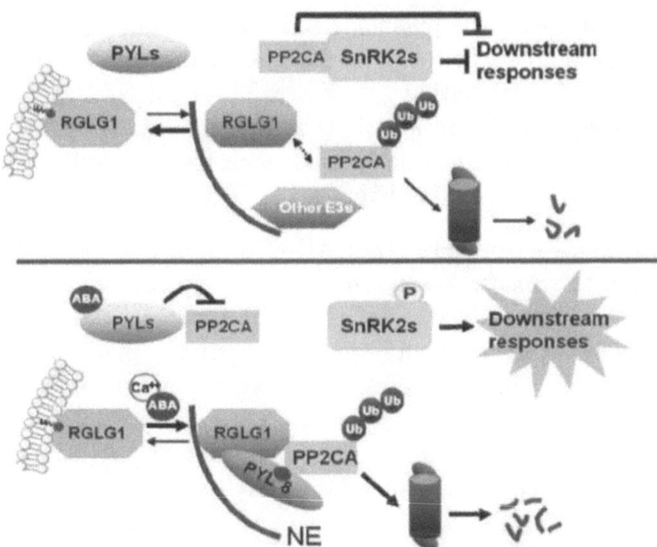

Figure 74. Proposed model for the enhanced degradation of PP2CA mediated by ABA, PYL8 and RGLG1. Under non-stress conditions (top, low ABA levels), RGLG1 shows less interaction with PP2CA (dashed arrow) because more protein is attached to the plasma membrane via N-myristoylated glycine 2 (purple circle). Other unidentified E3s might regulate PP2CA levels. PP2CA interacts with SnRK2s, which prevents their activation and leads to the inhibition of downstream ABA responses. PP2Cs themselves also inhibit downstream targets such as TFs and SLAC1. When plants are submitted to abiotic stress (bottom, high ABA levels), ABA promotes the interaction of certain PYLs with PP2CA, inhibiting its phosphatase activity. Additionally, ABA promotes the inhibition of RGLG1 myristoylation and cycloheximide-insensitive translocation to the nucleus (NE, nuclear envelope), where receptor–ABA–PP2CA complexes are formed. RGLG1 recognition of PP2CA is facilitated in the receptor–ABA–PP2CA complex.

some myristoylated proteins can dynamically relocalize in response to specific signals (Ding et al., 2019). Here, we showed that ABA is a signal that can mobilize myristoylated proteins by removing the myristoylation mark. This translocation facilitates the recruitment of the E3 in the nucleus, where PP2CA is predominantly localized. We also observed that the co-expression of RGLG1 with PP2CA increased the nuclear localization of RGLG1 to a level close to that observed for the RGLG1[G2A] mutant (**Figure 69**). The high nuclear intensity of RGLG1[G2A] is due to lack of capability to interact with the plasma membrane. We also showed that RGLG1 is able to interact with some ABA receptors (**Figure 66**), so the formation of the receptor–ABA–phosphatase complexes may facilitate their recognition by the E3 ligase, due to additional contact points being established with the receptors (**Figure 67** & **Figure 71**).

Data analysis revealed that ABA, drought and osmotic and oxidative stresses treatments diminishes the expression of *NMT1* (**Figure 58**). NMT1 is the central enzyme for myristoylation, which suggests that the enhanced mobilization of myristoylated proteins might be a plant response to cope with different stresses. NMT1 diminished expression may help to avoid myristoylation of newly synthesised proteins that and facilitate the function of putative demyristoylases (Boisson et al., 2003). However, no demyristoylation enzymes are known in plants. Alternatively, other mechanisms could deliver surface proteins to the nucleus, as nuclear envelope-associated endosomes can discharge their contents into the nuclear envelope and therefore transfer cargo from the cell surface into the nucleoplasm. Regarding myristoylation enzymes, there is also a NMT2 in *A. thaliana* regarded as a non-active enzyme because it has mutations in the active site. NMT1 OE lines or loss-of-function mutants show developmental defects. However, NMT2 is not able to complement *nmt1* mutants, which are lethal (Pierre et al., 2007). Future research is needed to investigate whether, in addition to the transcriptional downregulation of *NMT1* by ABA, there is an effect of ABA on the activity of NMT1 or how the myristoylome changes in response to ABA. It could be also interesting to study the activation of proteolytic processing at N-myristoylated glycine residues, as it has been described for the N-myristoylated glycine of ADP-ribosylation factor 1 by *Shigella flexneri* virulence factor IpaJ (**Figure 75**). IpaJ is a cysteine protease that catalyses the hydrolysis of N-myristoylated glycine at the amino terminus of host proteins (Burnaevskiy et al., 2013).

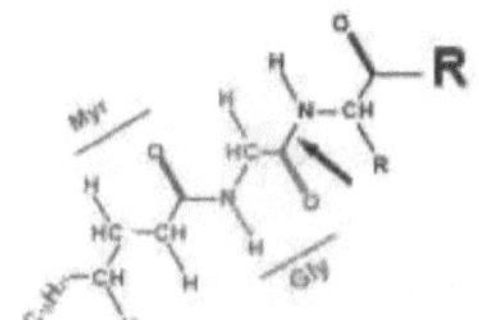

Figure 75. Target motif for IpaJ. Arrows indicate the protease cleavage sites. Glycines are coloured red and myristoyl group coloured blue. Reprinted from (Burnaevskiy et al., 2015).

In summary, we showed that hormone- and stress-induced shuttling of an E3 ligase is an important cellular mechanism to regulate hormonal signaling and cope with abiotic stress. Additionally, a better understanding of the dynamics of the RGLG1–PP2CA interaction is also provided in this work.

We have demonstrated that BPM3 and BPM5, substrate adaptors of the multimeric CRL3 E3 ligase, mediate recognition and ubiquitination of clade A PP2Cs, key repressors of ABA signaling (**Figure 76**). Data analysis in previous proteomic data obtained by Prof. Chengcai An's Laboratory, allowed us to find BPM3 and BPM5 as putative interactors of PP2CA (**Table 2**). In this proteomic study it was previously found RGLG1, which validates the experiment (Wu et al., 2016). However, this result needs to be validated by other techniques. We reported the interaction of both BPM3 and BPM5 with no only PP2CA, but also with HAB1, ABI1 and ABI2 by BiFC (**Figure 19**). Additionally, we performed several techniques to confirm these protein-protein interaction such as, Pull-down (**Figure 24**), Y2H (**Figure 26**), Split-LUC (**Figure 27**) and CoIP (**Figure 28**) to further test some of these interactions between BPM3/5 and clade A PP2Cs.

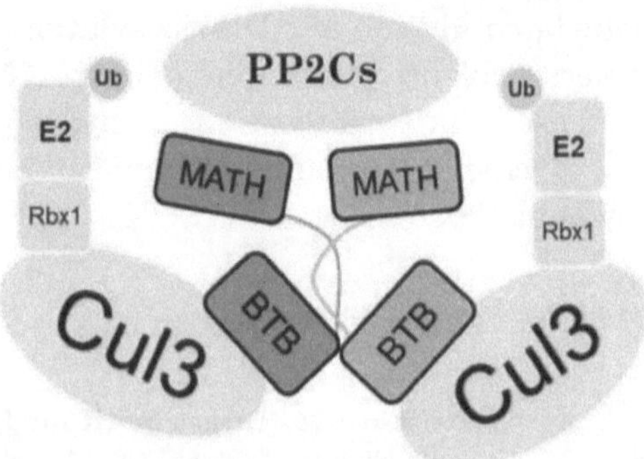

Figure 76. Model of CRL3[BPM] E3 ligase recognition of PP2Cs. Two BPM proteins can dimerize by their BTB domain to bring two CRL3 E3 ligases for the ubiquitination of one substrate, in this case a clade A PP2C.

We used specific antibodies against PP2CA, HAB1 and ABI1, which are fundamental tools to validate the degradation *in vivo* of these substrates by an E3 ligase. We reported an accumulation of PP2CA, HAB1 and ABI1 in the *bpm3 bpm5* double mutant plants compared to wild-type Col-0 (**Figure 35**). Accumulation of clade A PP2Cs in *bpm3 bpm5* plants led to reduced sensitivity to ABA-mediated inhibition of seedling establishment and root-growth compared to wild-type Col-0 (**Figure 46** & **Figure 47**). Moreover, *bpm3 bpm5* plants also displayed lower leaf temperature compared to wild-type Col-0 (**Figure 43**), which implies that there are more stomata open. *amiR-bpm* plants impaired in the transcript expression of BPM1/4/5/6 displayed lower leaf temperature compared to *bpm3 bpm5* double mutant and wild-type Col-0 (**Figure 43**). This suggest that other BPMs may have similar functions regulating ABA signaling. Moreover, *amiR-bpm* plants showed higher water-loss in detached leaves than wild-type

Col-0 (**Figure 42**) and impaired ABA-induced stomatal closure (**Figure 44**). However, ABA-induced stomatal closure was not impaired in *bpm3 bpm5* double mutant plants (**Figure 45**), which suggest certain functional redundancy of the BPM family in the control of stomatal aperture.

Increasing amounts of BPM3 and BPM5 led to reduced levels of phosphatases in transient expression in *N. benthamiana* plants (**Figure 29, Figure 30 & Figure 31**) and in *A. thaliana* OE lines (**Figure 33**) compared to wild-type Col-0. Reduced levels of clade A PP2Cs in BPM3 and BPM5 OE lines led to enhance ABA-mediated inhibition of seed germination, seedling establishment and root growth compared to wild-type Col-0 (**Figure 38**). Furthermore, we reported reduced water-loss in detached leaves of BPM3 and BPM5 OE lines compared to wild-type Col-0 (**Figure 39**), and consequently, BPM5 OE lines showed enhanced drought resistance compared to wild-type Col-0 (**Figure 40**). Finally, we demonstrated that ubiquitination of PP2CA was dependent on BPM function (**Figure 50 & Figure 51**).

BPM2 and BPM4 interact with DREB2A, which is a key transcription factor in both drought and heat stress tolerance (Morimoto et al., 2017). DREB2A induces the expression of many drought- and heat stress-inducible genes, so DREB2A degradation by BPMs would affect negatively the drought tolerance. However, BPM5 OE lines showed enhanced drought resistance compared to wild-type Col-0 (**Figure 40**). This suggests that different BPMs may have different targets or times of activation depending on environmental conditions, and that BPM5 may be more important to regulate drought stress responses. Other possibility is that the positive effect of PP2Cs degradation cancel out the negative effect of DREB2A degradation.

Additionally, PP2Cs are not the only substrate of the BPMs which degradation positive regulates the ABA signaling. ATHB6 is a transcription factor that negatively regulates a subset of ABA responses, such as ABA-mediated inhibition of seed germination and stomatal closure (Lechner et al., 2011). This adds importance to the regulation of the BPMs over the ABA signaling pathway. However, the regulation of the PP2Cs may be the key factor, as they are the main negative regulator of the pathway.

The degradation of PP2Cs by BPM3 and BPM5 was enhanced by ABA, this enables that at resting ABA levels PP2Cs protein levels remain sufficient to block ABA signaling. However, the mechanism whereby ABA enhances PP2Cs degradation via CRL3^{BPM} ubiquitination are still unknown and should be further investigated. In a previous Y2H assay it was found the interaction of CUL3 with BPM3 and BPM1, but not with BPM5 and BPM6 (Weber et al., 2005). This suggest that BPM5 may need some plant specific modification to be able to interact with CUL3, or that BPM5 is performing E3 activity without interaction with CUL3, maybe through other protein. The second option is less likely because BPM5 has not any domain to interact with an E2 or with other cullin. Moreover, BPMs can homo- and hetero-dimerize (Lechner et al., 2011). This open the possibility to speculate that BPM5 needs to dimerize with BPM3, or other BPM able to interact with CUL3, to function as a substrate adaptor of CRL3. Additionally, it is interesting that in all OE lines tested the amount of BPM5 protein was always higher than BPM3 (**Figure 22 & Figure 32**). If BPM3 is the rate limiting enzyme for the ubiquitination of PP2Cs, regulation of BPM3 would affect CRL3^{BPM} function while the regulation of BPM5 is not needed. Indeed, we reported BPM3 degradation by the 26S proteasome and its stabilization by ABA, while BPM5 is not being affected (**Figure 41**). As it has been reported for other substrate adaptors of CRLs, one possible way for this regulation would be the auto-ubiquitination of BPM3 attached to CUL3 without a substrate to ubiquitinate (Bosu & Kipreos, 2008). However, if BPM5 was completely dependent of BPM3 to function, the *bpm3* single mutant plants and the *bpm3 bpm5* double mutant plants should have the same phenotype. However, that is not the case. *bpm3* and *bpm5* single mutant plants didn't show reduced sensitivity to ABA compared to wild type Col0 (Lechner et al., 2011). While our *bpm3 bpm5* double mutant plants showed reduced sensitivity to ABA compared to wild type Col0 (**Figure 46 & Figure 47**). This suggest that BPM5 is able to function as a CRL3 substrate adaptor without the need of BPM3. This could be explained if BPM5 could interact with CUL3 through other BPMs, suggesting that more BPMs are involved in PP2Cs degradation, in agreement with the stronger phenotype of the *amiR-bpm* plants compared with the *bpm3 bpm5* double mutant plants (**Figure 42, Figure 43 & Figure 44**). Among them, BPM1 interacted with CUL3 in a Y2H assay (Weber et al., 2005), which makes BPM1 a good candidate.

In summary, we have discovered the first multimeric E3 ligase that targets PP2Cs, CRL3[BPM], and we have investigated the role of substrate adaptors BPM3 and BPM5 in PP2Cs ubiquitination. This represents a mechanism for nuclear degradation of clade A PP2Cs.

Finally, since several E3 ligases regulate PP2C levels, it might be reminiscent of the team tagging cooperation described in humans to regulate substrate ubiquitination by different E3-E3 pairs (Scott et al., 2016). Thus, exquisite regulation of substrate ubiquitination can be achieved by combination of different E3s in mammals, which is an interesting issue to be investigated in the plant field. The degradation of PP2Cs could be achieved through RGLG1/5 and CRL3[BPM] working together. To investigate this possibility, genetic analysis with BPM/RGLG knockdown plants will be required in future analyses.

CONCLUSIONS

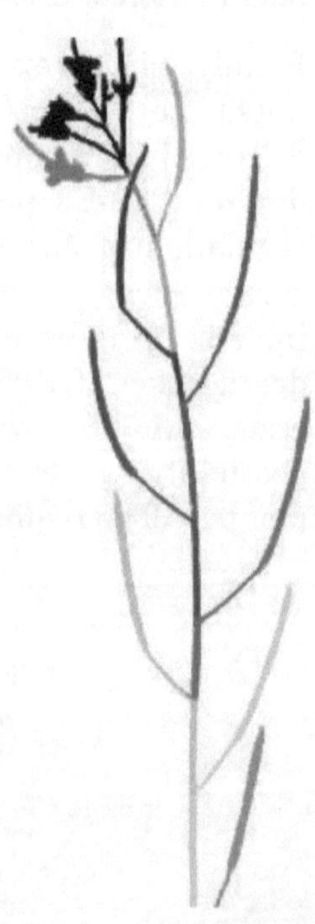

1. Using IP/MS data provided by Prof. Chengcai An's Laboratory, we discovered BPM3/4/5 as possible interactors of PP2CA.

2. We have confirmed nuclear interaction among BPM3/5 and different PP2Cs, i.e. PP2CA, HAB1, ABI1 and ABI2, using different techniques, i.e. BiFC, PD, CoIP, Split-LUC and Y2H.

3. BPM3/5 promote the degradation of PP2CA, HAB1 and ABI1 *in vivo*.

4. BPM3/5 OE lines showed enhanced ABA-mediated inhibition of seed germination, seedling establishment and root growth compared to wild-type Col-0, and also reduced water-loss in detached leaves. BPM5 OE lines showed enhanced drought resistance compared to wild-type Col-0.

5. Loss-of-function *bpm3 bpm5* double mutant plants showed reduced sensitivity to ABA-mediated inhibition of seedling establishment and root-growth compared to wild-type Col-0, and also lower leaf temperature.

6. *amiR-bpm* plants impaired in the transcript expression of BPM1/4/5/6 showed higher water-loss in detached leaves than wild-type Col-0 and displayed lower leaf temperature in intact plants.

7. ABA-induced stomatal closure is impaired in *amiR-bpm* plants, but not in *bpm3 bpm5* double mutant plants, which suggests certain functional redundancy of the BPM family in the control of stomatal aperture.

8. Ubiquitination of PP2CA is dependent on BPM3/5 function, leading presumably to its degradation via the 26S proteasome.

9. RGLG1 localizes mostly at the plasma membrane through a myristoylation modification under non-stress conditions.

10. RGLG1[G2A] mutant was not myristoylated *in* vivo and it was localized constitutively in the nucleus.

11. RGLG1 translocated to the nucleus after salt and osmotic stresses, and translocation after ABA treatment was CHX-resistant. Increased levels of the PP2CA target also induced RGLG1 relocalization to the nucleus.

12. NMT1 expression is downregulated after ABA treatment, osmotic and oxidative stresses and drought, inhibiting RGLG1 myristoylation.

13. RGLG1 is able to interact with some monomeric ABA receptors, i.e. PYL4, PYL8 and PYL9; forming a nuclear complex RGLG1/PP2CA/PYL8 in the presence of ABA.

MATERIALS AND METHODS

Biological materials

Bacterial Strains

· *Escherichia coli* DH5α and TOP10® strains (Invitrogen).

· *E. coli* DB3.1 strain (Life Technologies) for GATEWAY™ plasmids containing ccdB toxic gene.

· *E. coli* BL21 (DE3) pLysS strain for protein expression.

· Rosetta™ 2 (DE3) strain (Invitrogen), BL21 derivatives designed to enhance the expression of eukaryotic proteins that contain codons rarely used in *E. coli*.

· *Agrobacterium tumefaciens* C58C1 strain containing disarmed Ti plasmid pGV2260 (Deblaere et al., 1985).

· *A. tumefaciens* GV3101 strain (Lifeasible) containing disarmed nopaline type Ti plasmid pMP90.

Yeast Strain

· *Saccharomyces cerevisiae* AH109 strain (Clontech Laboratories).

Plant Material

· *A. thaliana* Col-0 ecotype.

· *N. benthamiana*.

· Mutants and transgenic stable lines in *A. thaliana* (**Table 3**).

Mutants and Transgenic Lines	Reference
hab1-1 (SALK_002104)	(Saez et al., 2004)
hab1-1 abi1-2 (SALK_002104; SALK_72009)	(Saez et al., 2006)
pp2ca-1 (SALK_028132)	(Kuhn et al., 2006)
bpm3 bpm5 (WiscDsLox239E10; SALK_038471C)	In this work
bpm3 bpm5 pp2ca1 ("")	In this work
amiR-bpm	(Lechner et al., 2011)
Pro35S: RGLG1-GFP (pBI121) (pMDC83)	(Wu et al., 2016) In this work
Pro35S: RGLG1[G2A]-GFP (pBI121) (pMDC83)	Chengcai An In this work
Pro35S: RGLG1-GFP / Pro35S: 3HA-PYR1	In this work
Pro35S: RGLG1-GFP / Pro35S: 3HA-PYL8	In this work
Pro35S: RGLG1-GFP / Pro35S: 3HA-PP2CA	In this work
Pro35S: 3HA-PP2CA	(Antoni et al., 2012)
Pro35S: HAB1	(Saez et al., 2004)
Pro35S: 3FLAG-PP2CA	(Wu et al., 2016)
Pro35S: 3HA-BPM3	In this work
Pro35S: 3HA-BPM5	In this work
Pro35S: GFP-BPM3	In this work
Pro35S: GFP-BPM5	In this work

Table 3. *A. thaliana* mutant and transgenic lines generated and used in this work.

Growing Conditions and Transformation

Bacterial Culture

Luria-Bertani media (LB; 10g/L triptone, 5g/L yeast extract and 10g/L NaCl, pH7.0 (2% agar in case of solid media preparation)) was used for bacterial culture. Different antibiotic-containing media was used depending on bacterial resistances (50µg/mL Kanamycin (Kan^R), 50µg/mL Spectinomycin (Spec^R), 100 µg/mL Ampicillin (Amp^R). The optimal temperature used for *E. coli* growth was 37°C and for *A. tumefaciens*, 28°C. In both cases, the liquid growth was performed in an orbital shaker at 250rpm.

Bacterial Transformation

For bacterial transformations, two techniques were used:

· Heat shock method:

100ng of plasmid DNA were incubated in 50µL of *E. coli* competent cells suspension for 30min on ice; the heat-shock was performed at 42°C for 45sec followed by 2min on ice. 500µL of S.O.C. Medium (Invitrogen) was added to the mixture and was incubated for 1h at 37°C. After, the mixture was spin-downed to collect the pellet and plated in LB medium containing the proper antibiotic.

· Electroporation method:

For *A. tumefaciens* 100ng of plasmid DNA were mixed with 50µL of competent cells suspension. The mixture was introduced in 0.2cm pre-cooled electroporation cuvettes (Bio-Rad). Electroporation procedure was performed in Eppendorf

Eporator®, 2000V/pulse (5-6ms) conditions. Immediately 1mL of LB medium was added and the mixture was incubated for 2h in an orbital agitator at 28°C. 200µL of the mixture was plated in LB medium supplemented with the proper antibiotic.

For *E. coli* the same amount of plasmid DNA was mixed with 50 µL of competent cells suspension. The mixture was introduced in 0.1cm pre-cooled electroporation cuvettes (Bio-Rad). Electroporation procedure was performed in Eppendorf Eporator®, 1800V/pulse (5-6ms) conditions. Immediately 500µL of S.O.C. Medium (Invitrogen) was added to the mixture and incubated for 1h at 37°C. After, the mixture was spin-downed to collect the pellet and plated in LB medium containing the proper antibiotic.

Yeast Culture

For *S. cerevisiae* culture, synthetic complete defined (SCD) culture medium (20% glucose, 7% yeast nitrogen base, 0.5M succinic acid pH5.5, 20x Drop Out solution (DO), 2% agar in case of solid media) and synthetic defined (SD) culture medium (20% glucose, 7% yeast nitrogen base (YNB), 0.5M succinic acid pH 5.5) were used.

For yeast transformants selection, SD basic culture medium was supplemented with extra amino acids depending on the autotrophy generated by the vector:

-Trp -Leu/SD: Nitrogen base supplemented with DO, His and Ade.

-Trp -Leu -His -Ade/SD: Nitrogen base supplemented with DO.

-Trp -Leu -Ade/SD: Nitrogen base supplemented with DO and His.

Yeast was grown at 28°C in an orbital shaker at 250rpm.

Protocol for generation of *S. cerevisiae* competent cells was done according to MATCHMAKER GAL4 Two-Hybrid System 3 Manual (Clontech Laboratories).

Yeast Co-Transformation

1μg of plasmid DNA from the two partners we are interested to test the interaction were mixed in an aliquot of AH109 yeast competent cells. 0.7mL of PEG-Li-TE Solution (4mL 45% PEG, 0.5mL 10x LiAc-TE, 0.5mL milli-Q water) was added. After vortex agitation, samples were incubated for 30min at 28°C. Vortex again was needed prior to next incubation for 20min at 42°C. Final centrifugation for 5min at 1000×g was used to collect the cells and resuspended in 100μL of milli-Q water. Glass beads were used to spread the culture on –Trp –Leu/SCD selective media. Plates were incubated for at least 2 days at 28°C.

Yeast Two-Hybrid (Y2H) Assays

Interaction assays were usually performed as described by Saez et al. (2008), using the AH109 yeast strain and testing yeast growth in medium lacking Histidine (-His) and Adenine (-Ade). The resulting transformants, in both cases, were grown O/N in liquid –Tryptophan (-Trp) –Leucine (-Leu)/SD culture medium and adjusted to equal cell density. Serial dilutions of cells were spotted on -Trp -Leu -His/SD and -Trp -Leu -His -Ade/SD culture medium.

Arabidopsis thaliana

In Vitro Tissue Culture

For *in vitro* growing assays, seed sterilization was performed using Sterilization Solution I (70% ethanol, 0.01% Triton X-100) for 10min followed by Sterilization solution II (50% Sodium hypochlorite) for 5 min. Removal of the sterilization solution II was done rinsing the seeds for 4 times with milli-Q water to fill the tube and changing the liquid with the pipette. Seeds were sown right after sterilization on Murashige–Skoog (MS) plates supplemented or not with different ABA concentrations per experiment. Stratification was conducted in the dark at 4°C for 3 days and after it, the plates were incubated in controlled-environment growth chamber at 22°C under 156 long day photoperiod conditions (16-h-light/8-h-dark photoperiod) at 80-100µE m^{-2} s^{-2}.

Plant Treatments

Treatments were performed differently depending on the nature of the media used:

ABA stock solution was prepared from solid (+)-ABA (SigmaAldrich). 26.4mg of powdered ABA were dissolved in 1mL of 50mM Tris-HCl pH8.0 buffer in order to get 10mM ABA stock solution.

· Solid media treatment:

MS plates were prepared with different concentrations of ABA. Germination and establishment experiments required ABA concentrations from 0.5 to 3µM. Root growth working concentration was 10µM ABA.

· Liquid culture treatments:

Analyses of protein expression under several drug treatments were done using liquid cultures. Plants were grown in liquid MS medium cultures and treated with different drugs. ABA treatment concentration was 50µM (+)-ABA prepared from stock solution of 10 mM ABA. MG-132 (UBPBio) working concentration was 50µM prepared by dilution from 5mM stock solution. CHX (Sigma-Aldrich) was used at 100µM from a stock solution of 100mM.

Greenhouse Culture

To propagate plants and for crosses, 1-week-old seedlings were transferred to soil (50% peat, 25% vermiculite, 25% perlite) and grown under long day conditions in the greenhouse 23°C/20°C temperature, 60% air relative humidity and 150µmol m−2 s−1 light.

Generation of Mutants

To obtain mutants in *A. thaliana*, parental plants were sowed on soil for 3 weeks. Closed flowers from mother plants were peeled to isolate the ovaries. Pollen from father plants was incorporated to the ovaries and the cross was kept in a plastic wrap for two days, allowing the cross to develop. Seeds generated from the cross, F1 seeds, were grown in MS plates under long day conditions. To get homozygous lines, selection was performed in F2 seeds.

Physiological Assays

Seed Germination and Seedling Establishment Assays

After surface sterilization of the seeds, approximately 100 seeds of each genotype were sown on MS plates supplemented or not with different ABA concentrations, 50mM NaCl or 200mM mannitol. Stratification was conducted in the dark at 4°C for 3 days. The plates were transferred to in vitro growing chamber under long day conditions. Radical emergence was analysed at 48-72h after sowing to score seed germination. Seedling establishment was scored as the percentage of seeds that developed green expanded cotyledons and the first pair of true leaves at 5-8 days.

Root Growth Assays

For root growth assays, seeds were sterilized and stratified as explained in a previous paragraph. Seedlings were grown on vertically oriented MS plates for 3-4 days. Afterwards, 20 plants were transferred to new MS plates lacking or supplemented with 10µM ABA. The plates were scanned on a flatbed scanner after 10 days to produce image files suitable for quantitative analysis of root growth using the NIH Image software ImageJ.

Water Loss and Drought Stress Experiments

For water loss analysis, plants were grown under greenhouse conditions (40-50% room humidity) and standard watering for 21d. 10 similar leaves per genotype were excised and submitted to the drying atmosphere of a laminar flow hood or the room ambient of the laboratory. Gravimetric analysis of water loss was performed and represented as the percentage of initial fresh weight loss at each scored time point.

For drought stress experiments, plants were grown under greenhouse conditions (20 individuals of each genetic background per experiment) and standard watering for 15d and then subjected to drought stress by stopping irrigation during 12d. Next, watering was resumed, and survival rate was calculated after 8d by counting the percentage of plants that had more than four green leaves. Photographs were taken at the start of the experiment (d-0), 12 days after stopping irrigation (d-12) and 8 days after rewatering (d-20).

Infrared Thermography

A. thaliana were grown in a controlled environment growth chamber at 22°C under a 12h light, 12h dark photoperiod at $100\mu E\ m^{-2}\ sec^{-1}$ and 40-50% room humidity. Philips bulbs were used (TL-D Super 8036W, white light 840, 4000K light code). Infrared thermographic images of rosette leaves were acquired from 6-week-old plants with a thermal camera FLIR E95 equipped with a 42° lens. Images were processed and quantified with the FLIR tools software. For quantification, the average temperature of 15 different sections corresponding to 4 leafs per plant were calculated. 5 plants per genotype were analysed in each experiment. The mean temperature ± standard deviation of all the plants for each genotype was reported. Statistical comparisons among genotypes were performed by pairwise t-tests.

Gas Exchange Experiments

This protocol was performed by Dr. Ebe Merilo's laboratory. For gas exchange experiments, *A. thaliana* seeds were planted in soil containing 4:3 (v:v) peat:vermiculite and grown in pots either well-watered or water-stressed. Soil water holding capacity of well-watered pots was 95±3% of the absolute water holding capacity calculated based on dry soil weight, whereas that of drought pots was 36±3% during the gas exchange measurements. Plants were grown in growth chambers (Snijders Scientific, Drogenbos, Belgia) at 12/12 photoperiod, 23/18°C temperature, 150µmol m^{-2} s^{-1} light and 70% relative humidity, and were 23-29 days old during experiments. Whole-rosette leaf conductances were recorded with an 8-chamber custom-built temperature-controlled gas-exchange device. Plants were inserted into the measurement cuvettes and allowed to stabilize at standard conditions: ambient CO_2 (~400 ppm), light 160µmol m^{-2} s^{-1}, relative air humidity ~63 ± 3%. Then, 10µM ABA with 0.012% Silwet L-77 (Duchefa) and 0.05% ethanol was sprayed on the leaves, plants were put back into cuvettes and measurements of leaf conductance continued. Photographs of plants were taken after the experiment and leaf rosette area was calculated using NIH Image software ImageJ. Leaf conductance for water vapour was calculated with a custom written program (Kollist et al., 2007). To calculate Gst (stomatal conductance), cuticular conductance was determined for studied lines using an excision method (Jakobson et al., 2016) and used to calculate Gst from leaf conductance.

Nucleic Acids Extraction and Analysis Methods

DNA Extraction

Escherichia coli

Plasmid DNA was extracted from the bacteria using alkaline lysis of the cells. 1.5mL of saturated cultures of *E. coli* were centrifuged at 12000×g for 1min. Supernatant was discarded and the pellet was resuspended in 100μL of ultrapure water of Type 1 (milli-Q water) and 100μL of Lysis Solution (0.1M NaOH, 10mM ethylene diamine tetraacetic acid (EDTA), 2% sodium dodecyl sulfate (SDS). Samples were heated at 95°C for 2 min and next kept on ice. 50μL of 1M $MgCl_2$ were added and after vortex the solution, the tubes were centrifuged at 12000×g for 5min. Neutralization was performed with 50μL of 5M AcK (balanced with acetic acid at pH 5.0). The plasmid DNA was precipitated with 2 volumes of pre-cooled 96% ethanol. After 15min of incubation on ice, the precipitate was collected by centrifugation at 12000×g for 15min. After discarding the supernatant, the pellet was rinsed with 500μL of 70% ethanol and centrifuged at 12000×g for 5min. The pellet was air-dried and resuspended with 30μL of milli-Q water.

Arabidopsis thaliana

For genomic DNA extraction, 100mg of leaf material from 2-weeks-old plants was collected in tubes and freeze into liquid nitrogen. The material was grinded with a glass pistil until fine powder. 2 volumes of Extraction Buffer (2% cetyl trimethyl ammonium bromide (CTAB), 100 mM Tris-HCl pH8.0, 20mM EDTA, 1.4 M NaCl) were added to the sample and incubated at 65°C for 10 min. 1 volume of chloroform/isoamyl alcohol (24:1) was added next and after vortexing, the sample was centrifuged at 12.000×g for 10min. 1 volume of the aqueous phase was transferred to a new tube and 1/10 of 10% CTAB was added (10% CTAB is very viscous so prewarming of the solution at 65°C has to be done prior to use) followed by an incubation at 65°C for 2min. For induction of the CTAB/DNA-RNA complex precipitation, 2 volumes of milli-Q water were added and followed by 15min of ice incubation. Centrifugation of 12000×g for 10min was performed to collect the pellet. After discarding the supernatant, 400µL of 1M NaCl was used to resuspend the pellet. To precipitate de DNA, 800µL of 96% ethanol were added followed by incubation on ice for 15min. Last step was collecting the genomic DNA by centrifugation at 12.000×g for 15min. Samples were rinsed with 70% ethanol, as described before and the genomic DNA was resuspended with 30µL of milli-Q water.

Gene Expression Analysis by PCR Reaction

For cloning of open reading frame (ORF) amplifications and expression analysis by PCR specific pairs of primers were used for each gene (**Table 4**).

Amplification	Primer name	Sequence (5'-3')
MATH3 (1-529 aa)	F_{ATG} BPM3.1	ACCATGGGTACCGTCGGAGGTATAG
	R_{LVI} BPM3	CTATAGCACAATGCCATACTGTTTTGG
MATH5 (1-537 aa)	F_{ATG} BPM5	ACCATGGCAGAATCAGTG
	R_{VHV} BPM5	CTAGACGTGAACAGAGTGTAACTG
BTB3 (450-1.227 aa)	F_{CLV} BPM3	TGTCTTGTCATCAATTGTACTG
	R_{Stop} BPM3	CTAAGACACTGCTCGCACTTCTTTTCG

FLAG-Ub (1-282 aa)	FwNco FLAGUB	ACCATGGCATACCCATACGACGTTC
	RvStop FLAGUB	TCAACCACCACGGAGCCTGAGGACC

Table 4. Primer sequences for amplification of ORFs. Stop codons are highlighted in red. In brackets are the amplification size measured in aa.

For genotyping the mutants, CTAB protocol for genomic DNA extraction was performed and Taq polymerase (produced in our laboratory) was used for PCR analysis. Pairs of primers used for genotyping each mutant and insertions (SALK and WiscDsLox) are specific for each gene (**Table 5**).

Mutant	Primer name	Sequence (5'-3')
bpm3 (genomic)	F_{CLV} BPM3	TGTCTTGTCATCAATTGTACTG
	R_{Stop} BPM3	CTAAGACACTGCTCGCACTTCTTTTCG
bpm3 (T-DNA)	F_{ATG} BPM3	ATGAGTACCGTCGGAGGTATAG
	R_{p745}	AACGTCCGCAATGTGTTATTAAGTTGTC
bpm5 (genomic)	F_{ATG} BPM5	ACCATGGCAGAATCAGTG
	R_{PAT} BPM5	GTTGCAGGCTCCACATCTTCCGG
bpm5 (T-DNA)	F_{ATG} BPM5	ACCATGGCAGAATCAGTG
	R_{pROK2}	GCCGATTTCGGAACCACCATC
pp2ca-1 (genomic)	F_{PP2CA}	ACCATGGCTGGGATTTGTTGCGGTG
	R_{PP2CA}	GTCGACAGACGACGCTTGATTATTCCTC
pp2ca-1 (T-DNA)	F_{PP2CA}	ACCATGGCTGGGATTTGTTGCGGTG
	R_{pROK2}	GCCGATTTCGGAACCACCATC

Table 5. Primer sequences for genotyping double/triple mutants generated in this work.

RNA extraction

RNA was extracted from seedlings using E.Z.N.A. Plant RNA kit (Omega, R6827-01), following the manufacturer's instructions. cDNA was synthesized from 2 µg of total purified RNA using 30U of RevertAid Reverse Transcriptase (Thermo Scientific).

RT-qPCR

RT-qPCR was performed using PyroTaq EvaGreen qPCR Master Mix 5X (Cultek), which includes EvaGreen® Dye and carboxy-X-rhodamine (ROX) as a passive reference dye. The reaction was performed in a final volume of 10µL using 0.4µL of cDNA. Specific primer pairs for each gene were used in this analysis (**Table 6**). qPCR was performed in the 7500 Fast Real-Time PCR System (Applied Biosystems). Relative quantification of gene expression data was carried out using the $2^{-\Delta\Delta CT}$ or comparative CT method (Livak & Schmittgen, 2001). Expression levels were normalized using the CT values obtained for the *ACTIN8* gene. The presence of a single PCR product was further verified by dissociation analysis in all amplifications.

Gene	Primer name	Sequence (5'-3')
ACTIN8	F_{ACT8}	AGTGGTCGTACAACCGGTATTGT
	R_{ACT8}	GAGGATAGCATGTGGAAGTGAGAA
RAB18	F_{RAB18}	TGGCTTGGGAGGAATGCTTCA
	R_{RAB18}	CCATCGCTTGAGCTTGACCAGA
RD29B	F_{RD29B}	CCATCGCTTGAGCTTGACCAGA
	R_{RD29B}	TCAGTTCCCAGAATCTTGAACT
KIN10	F_{KIN10}	TCAGTTCCCAGAATCTTGAACT
	R_{KIN10}	TTCCCGCCTGTTGTGCTC

RD22	F$_{RD22}$	TTCCCGCCTGTTGTGCTC
	R$_{RD22}$	TGGCAGTAGAACACCGCGAAT
BPM3	F$_{LKA}$ BPM3	GCTCAAAGCGTTCTGCTTAGA
	R$_{Stop}$ BPM3	CTAAGACACTGCTCGCACTTCTTTTCG
BPM5	F$_{KEL}$ BPM5	AAGGAACTAAAAGGCGTTTGC
	R$_{VSQ}$ BPM5	ACACGCTCTGAGACTTTGCTC

Table 6. Primers sequences for RT-PCR analysis.

Generation of new constructs

GATEWAY™ Cloning

The GATEWAY™ system pCR®8/GW/TOPO® TA Cloning® Kit (Life Technologies) was used for generating destiny vectors following manufacturer's instructions. BPM3 and BPM5 were cloned from the pENTR221, all the other constructions were cloned from pCR8.

50-150ng of entry vector DNA and 150ng of destiny vector DNA were mixed with 1.5µL LR Clonase™ II enzyme mix (Life Technologies). Volume was brought to 7.5µL with milli-Q water and incubated at 25°C for 1h. To inactivate the clonase enzyme, 1µL of 1% Proteinase K solution was added to the sample and incubated at 37°C for 10min. 5µL of reaction was transformed in a 50µL aliquot of DH5α *E. coli* chemocompetent cells.

Constructions for Yeast-Two Hybrid

For Y2H assays, pGADT7-GW and pGBKT7-GW versions and pGADT7 and pGBKT7 in their restriction versions were used. The genes cloned in these vectors (**Table 7**) were fused to the GAL4 activation domain in pGADT7 or to the GAL4 binding domain in pGBKT7. We used the GATEWAY™ system for the generation of new constructs.

Constructions (Y2H)	Plasmid	Bacterial selection	References
GAD-PP2CA	pGADT7	Amp^R	(Fujii et al., 2009)
GAD-ABI1	pGADT7	Amp^R	(Fujii et al., 2009)
GAD-HAB1	pGADT7	Amp^R	(Santiago et al., 2009)
GAD-HA-RGLG1	pGADT7 GW	Amp^R	In this work
GAD-HA-RGLG1G2A	pGADT7 GW	Amp^R	In this work
GBD-myc-BPM3	pGBKT7 GW	Kan^R	In this work
GBD-myc-BPM5	pGBKT7 GW	Kan^R	In this work
GBD-PYL4	pGBKT7	Kan^R	(Santiago et al., 2009)
GBD-PYL8	pGBKT7	Kan^R	(Santiago et al., 2009)
GBD-PYL9	pGBKT7	Kan^R	(Irigoyen et al., 2014)

Table 7. Constructs used in this work for Y2H assays.

Constructions for Plant Transformation

Constructions were both used for transient expression in *N. benthamiana* and stable transformation in *A. thaliana* (**Table 8**). We used the GATEWAY™ system for the generation of new constructs.

Constructions	Plasmid	Bacterial selection	References
35S:RGL1-GFP	pBI121	Kan[R]	(Wu et al., 2016)
35S:RGL1[G2A]-GFP	pBI121	Kan[R]	Chengcai An
2x35S:35S:RGL1-GFP	pMDC83	Kan[R]	In this work
2x35S:RGL1[G2A]-GFP	pMDC83	Kan[R]	In this work
35S:RGLG1-RFP	pGWB554	Kan[R]	In this work
2x35S:PP2CA-GFP	pMDC83	Kan[R]	(Antoni et al., 2012)
2x35S:GFP-ABI1	pMDC43	Kan[R]	In this work
2x35S:GFP-HAB1	pMDC43	Kan[R]	Américo Rodrigues
NOS:OFP-TM23	pGPTVII	Kan[R]	(Batistic et al., 2012)
35S:AtFib2-mRFP	pKT7	Kan[R]	(Herranz et al., 2012)
2x35S:GFP-BPM3	pMDC43	Kan[R]	In this work
2x35S:GFP BPM5	pMDC43	Kan[R]	In this work
2x35S:HA-BPM3	pAlligator2	Spect[R]	In this work
2x35S:HA-BPM5	pAlligator2	Spect[R]	In this work
35S:FLAG-Ub	pER8	Spect[R]	In this work

Table 8. Constructions used in this work for plant transformation.

Constructions for Bi-molecular Fluorescence Complementation (BiFC) and Multicolour BiFC

For BiFC experiments, destiny vectors for protein expression were generated (**Table 9**). The pSPYNE-35S and pSPYCE-35S vectors used in this work were obtained from (Walter et al., 2004) and pYFPN43 and pYFPC43 vectors were kindly provided by Alejandro Ferrando (Belda-Palazon et al., 2012). The proteins of interest were fused to N-terminal or C-terminal part of the yellow fluorescent protein (YFP) under 35S promoter control. For multicolour BiFC the vectors used were obtained from (Gehl et al., 2009). All these constructions have a 35S promoter. We used the GATEWAY™ system for the generation of new constructs.

Constructions (BiFC)	Plasmid	Bacterial selection	References
YFP^C-BPM3	pYFC43	Kan^R	In this work
YFP^C-BPM5	pYFC43	Kan^R	In this work
YFP^C-MATH3	pYFC43	Kan^R	In this work
YFP^C-BTB3	pYFC43	Kan^R	In this work
YFP^C-OST1$^{\Delta280}$	pYFC43	Kan^R	(Vlad et al., 2009)
YFP^N-ABI1	pYFN43	Kan^R	Marta Peirats-Llobet
YFP^N-ABI2	pYFN43	Kan^R	Américo Rodrigues
YFP^N-ΔPP2CA	pYFN43	Kan^R	(Saez et al., 2008)
YFP^N-PYL8	pYFN43	Kan^R	(Rodriguez et al., 2014)
HAB1-myc-YFP^N	pSPYNE	Kan^R	(Pizzio et al., 2013)
PP2CA-myc-YFP^N	pSPYNE	Kan^R	(Pizzio et al., 2013)
Constructions (mcBiFC)	Plasmid	Bacterial selection	References
VENUS^N-c-myc-PP2CA	pDEST–VYNE(R)^{GW}	Kan^R	Marta Peirats-Llobet
SCFP^N-FLAG-PYL8	pDEST–SCYNE(R)^{GW}	Kan^R	In this work
RGLG1-HA-SCFP^C	pDEST–^{GW}SCYCE	Kan^R	In this work

Table 9. Constructions used in this work for BiFC/mcBiFC experiments in *N. benthamiana*.

Constructions for Split-Luciferase

For Split-LUC experiments, destiny vectors for protein expression were generated (**Table 10**). All these constructions have a 35S promoter. We used the GATEWAY™ system for the generation of new constructs.

Constructions	Plasmid	Bacterial selection	References
BPM3-cLUC	pDEST-GWCLUC	Kan^R	In this work
BPM5- cLUC	pDEST-GWCLUC	Kan^R	In this work
PP2CA-nLUC	pDEST-GWNLUC	Kan^R	In this work
HAB1-nLUC	pDEST-GWNLUC	Kan^R	In this work

Table 10. Constructions used in this work for Split-LUC experiments in *N. benthamiana*.

Constructions used for Protoplast Transformation

For protoplast transformation, destiny vectors were generated (**Table 11**). We used the GATEWAY™ system for the generation of new constructs. Some constructions were kindly provided by Jörg Kudla's Laboratory.

Constructions	Plasmid	Bacterial selection	References
pRD29B::LUC	pSK	Amp^R	Jörg Kudla
35S::GUS	pSK	Amp^R	Jörg Kudla
35S::ABF2	pXCS GW	Amp^R	In this work
35S::OST1	pXCS GW	Amp^R	In this work

Table 11. Constructions used in this work for protoplast transformation.

Constructions for Protein Purification in *E. coli*

For expression of recombinant proteins, two systems were used: Histidine (His) purification and GST purification (**Table 12**). To generate 6His-MATH3 and 6His-MATH5, we cloned the corresponding NcoI-EcoRI fragment into pETM11 plasmid.

Constructions	Plasmid	Bacterial selection	References
GST-BPM5	pDEST15	Amp[R]	(Lechner et al., 2011)
GST-PP2CA	pGEXT4	Amp[R]	Lesia Rodriguez
His-PP2CA	pET28a	Kan[R]	(Antoni et al., 2012)
His-ΔNPP2CA	pET28a	Kan[R]	Regina Antoni
His-HAB1	pETM11	Kan[R]	Regina Antoni
His-ABI1	pCOLA Duet	Kan[R]	Regina Antoni
His-ABI2	pCOLA Duet	Kan[R]	Regina Antoni
His-MATH3	pETM11	Kan[R]	In this work
His-MATH5	pETM11	Kan[R]	In this work

Table 12. Constructions used in this work for protein purification in *E. coli*.

Protein Technology

His-tag Proteins Culture and Purification

BL21 (DE3) pLysS cells transformed with the corresponding construct were grown in 50ml of LB medium containing 50µg/ml kanamycin to an OD at 600 nm of 0.6-0.8. Then, 1mM IPTG was added and the cells were harvested 3h after induction at 37°C and 150rpm shaking and stored at -80°C before purification. The protein pellet was resuspended in 2 mL of HIS buffer (50mM Tris-HCl pH 7.6, 250mM KCl, 10%

glycerol, 0.1% Tween 20, and 10mM ß-mercaptoethanol), and the cells were sonicated with 2 pulses of 30s (hold position, 50% of the power) in a sonicator (Dr. Hielscher, UP200s). A cleared lysate was obtained after centrifugation at 14.000×g for 15 min at 4°C, and it was diluted with 2 volumes of HIS buffer. The protein extract was applied to a 0.5mL Ni-NTA acid agarose column, and the column was washed with 10 mL of HIS buffer supplemented with 20% glycerol and 30 mM imidazole. Bounded protein was eluted with HIS buffer supplemented with 20% glycerol and 250mM imidazole.

Recovery of the columns was done by adding 5 mL of 0.2 M acetic solution supplemented with 30% glycerol, washed with 8 mL of milli-Q water and the resin was kept at 4°C with 5 mL of 30% ethanol.

GST-tag Proteins Culture and Purification

BL21 (DE3) pLysS cells transformed with the corresponding construct were grown in 50ml of LB medium containing 50µg/ml ampicillin to an OD at 600 nm of 0.6-0.8. Then, 1mM IPTG for GST-PP2CA and 0.1mM IPTG for GST-BPM5 was added and the cells were harvested 4h after induction for GST-BPM5 and after O/N for GST-PP2CA, both at 16°C and 150rpm shaking, and stored at -80°C before purification. The protein pellet was resuspended in 2 mL of sonication buffer (1xTBS, 1mM EDTA, 0,1% Triton X-100 and 0,1% Tween-20), and the cells were sonicated with 4 pulses of 30s (hold position, 50% of the power) in a sonicator (Dr. Hielscher, UP200s). A cleared lysate was obtained after centrifugation at 15.000×g for 15 min at 4°C, and it was diluted to 6mL with sonication buffer. The protein extract was applied to a 0.6mL Glutathione Sepharose 4B column, and the column was washed with 10 mL of TBS. Bounded protein was eluted with TBS supplemented 10mM reduced glutathione.

Recovery of the columns was done by adding 5 mL of regeneration buffer 1 (0,1M Tris-HCl, 0,5M NaCl, ph8,5), 5mL of regeneration buffer 2 (0,1M sodium acetate, 0,5M NaCl, pH 4,5), repeated both previous steps and washed with 5 mL of TBS. The resin was kept at 4°C with 5 mL of 20% ethanol.

Coomassie Staining

For protein visualization, acrylamide gels were incubated in InstantBlue™ (Expedeon) staining solution for 15 min. Unstaining of the gel was done rinsing the gel with milli-Q water until the background was removed.

Transient Protein Expression in *N. benthamiana*

A. tumefaciens expressing our constructs were grown in liquid LB medium to late exponential phase and cells were harvested by centrifugation and resuspended in 10mM MES-KOH pH 5.6, containing 10mM MgCl2 and 150µM acetosyringone to an OD600 of 1. These cells were mixed with an equal volume of *A. tumefaciens* expressing the silencing suppressor p19 of Tomato bushy stunt virus so that the final density of *A. tumefaciens* solution was ~1. Bacteria were incubated for 3h at room temperature and then infiltrated into young fully expanded leaves of 4-week-old *N. benthamiana* plants. Samples for BiFC or localization assays were examined after 3 to 4d with a Leica TCS-SL confocal microscope and a laser scanning confocal imaging system. Samples for immunoblot and immunoprecipitation assays were harvested, frozen in liquid nitrogen, and stored at −80°C.

Protein Expression in Protoplasts

A. thaliana protoplasts prepared from wild type Col-0 or *bpm3 bpm5* double mutant plants following the protocol from (Yoo et al., 2007) were transfected with the reporter construct pRD29B::LUC, p35S::GUS for normalization and either ABF2 or ABF2+OST1 expression cassettes. Protoplast suspensions were incubated for 6 h after transfection in the absence or presence of 5µM exogenous ABA added 3h after transfection. Samples were harvested, frozen in liquid nitrogen and stored at -80°C. To measure the luminescence activity, samples were resuspended in lysis protoplasts buffer (25mM Tris-HCl pH8, 1mM DTT, 2mM DACTAA, 10% Glycerol and 1% Triton X-100). The luminescence activity of the LUC reporter was measured with the help of the Luciferase Assay System (Promega, E1500) in a luminometer Glomax Multi Detection System (Promega, E7071) according to the manufacturer's instructions, and normalized with the fluorescence signal of the GUS reporter after adding 50µL of GUS reveal Buffer (0,035g MUG, 10mM Tris-HCl pH8, 2mM MgCl2 and H_2O to 100mL) for 20µL of sample and measured in a fluorimeter Glomax Multi Detection System (Promega, E7071).

Protein Extraction from Plant Material

For protein extraction from both *A. thaliana* and *N. benthamiana* material, 2 volumes of Lysis Buffer (50mM Tris-HCl pH8.0, 150mM NaCl, 1% Triton X-100, 3mM DTT, 50µM proteasome inhibitor MG-132 and 1 tablet of protease inhibitor cocktail mini EDTA free Roche (1 tablet/10mL buffer)) were added to the sample. The mixture was kept on ice and vortexed. Immediately, samples were centrifuged at 12.000×g at 4°C for 30 min. The supernatant was transferred into a new tube and the protein quantification was performed by Bradford assay (Bio-Rad).

SDS-PAGE Electrophoresis

Mixed samples with Laemmli buffer were boiled for 10min at 95°C and spin-down. The system used for protein analysis was MiniProtean® System from Bio-Rad. Running gel containing 8-12% Acrylamide (19:1 Acrylamide/Bis-acrylamide, National diagnostics), 375mM Tris-HCl pH8.8, 0.1% SDS, 0.2% N,N,N',N'-tetramethyl ethylenediamine (TEMED) and 0.08% ammonium persulfate (APS). Stacking gel is composed by 4% acrylamide, 125mM Tris-HCl pH6.8, 0.1% SDS, 0.8% TEMED and 0.1% APS.

Western Blot Analyses

After SDS-PAGE, wet transfer method was used for protein visualization. Proteins were transferred to a polyvinylidene difluoride (PVDF) membrane Immobilon®-P (Millipore™), previously activated in 100% methanol solution, using Mini Trans-Blot® Cell system (Bio-Rad) at 4°C and 110V during 1.5h. Transfer buffer used was 1x Towbin Buffer (25mM Tris-HCl pH7.6, 192 mM glycine, 20% (v/v) methanol, 0.1% SDS).

To check the transference, the membrane was incubated in Ponceau S Solution (0.1% (w/v) Ponceau S (Sigma-Aldrich) in 5% acetic acid) for 15 min in an orbital shaker. 1% acetic acid was used for membrane unstaining and 1x Tris-buffered saline (TBS, 50mM Tris-HCl pH 7.6, 150mM NaCl) was used to remove completely the staining.

For protein detection, the membrane was incubated at room temperature for at least 2 h in blocking solution (1x TBS, 0.1% Tween-20 with 5% (w/v) non-fat dry milk). The antibodies (**Table 13**) were incubated at least for 1h at room temperature in 5% blocking solution. After the antibody incubation, 3 washes with 1x TBST for 10min were performed to remove the excess of primary antibody, and next, secondary antibody which was also diluted in 5% blocking solution was incubated for 1 hour and 3 washes were done as previously described. Detection was performed using the ECL advance western blotting chemiluminescent detection kit (GE Healthcare).

Image capture was done using the image analyser LAS3000, and quantification of the protein signal was done using NIH Image software ImageJ.

Antibodies

To obtain specific antibodies against HAB1, recombinant GST-HAB1 protein was prepared as previously described (Park et al., 2009) and injected with complete Freund's adjuvant into two rabbits, and boosted twice at 3-week intervals by contractor (R.B. Sargeant, Ramona, CA). The quality of antiserum after the third boost was tested by immunoblot. Anti-HAB1 antibody was purified from the antiserum using His-HAB1 coupled to NHS activated agarose beads according to the manufacturer's instructions (Thermo Fisher).

The specific antibodies for PP2CA and ABI1 have been described previously (Kong et al., 2015; Wu et al., 2016) and the rest were bought to the corresponding manufacturer **(Table 13)**.

Primary antibodies	Description	Reference	Dilution
Anti-PP2CA	Polyclonal	(Wu et al., 2016)	1:2.000
Anti-ABI1	Polyclonal	(Kong et al., 2015)	1:2.000
Anti-HAB1	Polyclonal	This work	1:1.000
Anti-GFP (JL8)	Monoclonal	TaKaRa, 632381	1:10.000
Anti-HA/HRP	Monoclonal	Roche, 3F10	1:2.000
Anti-His	Monoclonal	Roche, BMG-his	1:2.000
Anti-GST	Polyclonal	Sigma, G7781	1:10.000

Anti-Actin	Polyclonal	Agrisera, AS13 2640	1:5.000
Anti-RFP (Biotin)	Polyclonal	Abcam, ab3477	1:10.000
Anti-FLAG	Monoclonal	Sigma, F3165	1:5.000
Anti-Ub (P4D1)	Monoclonal	Santa Cruz Biotechnology, sc-8017	1:10.000
Anti-GFP$_{Nt}$	Polyclonal	Sigma, G1544	1:2.000
Anti-Myristic acid	Polyclonal	Abcam, ab37027	1:1.000
Secondary antibodies	**Description**	**Reference**	**Dilution**
Anti-IgG (mouse)-HRP	Polyclonal	Abcam, ab205719	1:5.000
Anti-IgG (rabbit)-HRP	Polyclonal	Abcam, ab205718	1:5.000
Anti-IgG (rat)-HRP	Polyclonal	Thermo Fisher, A18739	1:10.000

Table 13. Primary and secondary antibodies used in this work for chemiluminescent detection of the proteins of interest. HRP (horseradish peroxidase).

Pull-down Assays

For pull-down assays, GST proteins were purified from a pellet using glutathione affinity chromatography as described previously. The beads were recovered in 3mL TBS and 50µg of the indicated His-tagged proteins were added. The mix was incubated 3h at 4°C with constant rocking. Afterwards, the proteins that didn't interact were washed away with TBS and the rest were eluted with 2X Laemmli buffer (125mM Tris-HCl pH 6.8, 4% SDS, 20% glycerol, 2% mercaptoethanol, 0.001% bromophenol blue). Finally, extracts were analysed by SDS-PAGE, followed by western blotting and immunodetection using anti-HIS antibodies or anti-GST antibodies.

Coimmunoprecipitation (coIP)

Co-IP experiments of PP2CA, HAB1 and ABI1 with BPM3/BPM5 were performed using agroinfiltration in *N. benthamiana*. Protein extracts were prepared in lysis buffer (50mM Tris-HCl pH8.0, 150mM NaCl, 1% Triton X-100, 3mM DTT, 50µM proteasome inhibitor MG-132 and 1 tablet of protease inhibitor cocktail mini EDTA free Roche (1 tablet/10mL buffer)) from *N. benthamiana* leaves 72h after agroinfiltration with constructs to co-express GFP or PP2C-GFP proteins and either HA-tagged BPM3 or BPM5. GFP or PP2C-GFP proteins were immunoprecipitated using super-paramagnetic micro MACS beads coupled to monoclonal anti-GFP antibody according to the manufacturer´s instructions (Miltenyi Biotec). Purified immunocomplexes were eluted in 2xLaemmli buffer, boiled and run in a 10% SDS-PAGE gel. Proteins immunoprecipitated with beads coupled to monoclonal anti-GFP antibody were transferred onto Immobilon-P membranes (Millipore) and probed with anti-HA to detect coIP of HA-tagged BPM3 or BPM5.

Split-luciferase (LUC) Complementation Assay

Split-LUC complementation assay was performed by transient expression in *N. benthamiana* leaves by agroinfiltration as described above but in this case the final density of the *A. tumefaciens* solution was 0.1 instead of 1. MG-132 (50µM) was applied into the infiltrated region 12h before inspection, which was performed 60h after infiltration. To this end, leaves co-expressing different constructs were examined for LUC activity by applying 1mM D-luciferin and placed in the dark for 5min before imaging. LUC complementation was observed with a CCD imaging system (LAS3000, Fujifilm) using 10min exposures.

For quantification of LUC activity, leaf samples were collected, ground in liquid nitrogen and immediately placed in lysis buffer (50mM Tris-HCl pH8.0, 150mM NaCl, 1% Triton X-100, 3mM DTT, 50μM proteasome inhibitor MG-132 and 1 tablet of protease inhibitor cocktail mini EDTA free Roche (1 tablet/10mL buffer)) in ice for protein extraction. Homogenates were cleared by centrifugation at 12.000 g, 4°C for 15min, and supernatants were used to quantify luminescence activity by the Luciferase Assay System (Promega, E1500) using 10 μg of protein. Luminescence was analyzed using a Glomax Multi Detection System (Promega, E7071). In order to normalize LUC values, we co-infiltrated all leaves with a GUS expression vector (pEXP::GUS) and GUS activity was analysed using 4-methyl umbelliferyl glucuronide (MUG) as substrate and the above mentioned detection system.

In vivo Protein Degradation Assays

For *in vivo* protein degradation experiments, *A. tumefaciens* cultures containing constructs that express HA-BPM3 or HA-BPM5, PP2CA-GFP and the silencing suppressor p19 were co-infiltrated at different ratios (0, ¼, 1 BPMs, while PP2CA remained constant) in *N. benthamiana* leaves. Three days after infiltration, samples were collected, ground in liquid nitrogen and immediately placed in lysis buffer (50mM Tris-HCl pH8.0, 150mM NaCl, 1% Triton X-100, 3mM DTT, 50μM proteasome inhibitor MG-132 and 1 tablet of protease inhibitor cocktail mini EDTA free Roche (1 tablet/10mL buffer)) on ice for protein extraction. Homogenates were cleared by centrifugation at 12.000 g, 4°C for 15 min, and supernatants were used for protein immunoblot analysis. Samples were also collected for *Actin* and *PP2CA* mRNA analyses to ensure equal amounts of *PP2CA* transcript were expressed in different co-infiltrations. Similar methods were used for the analysis of GFP-ABI1 and GFP-HAB1 degradation promoted by HA-BPM3 or HA-BPM5.

Protein Stability Kinetics

Seedlings of wild type Col-0 or *bpm3 bpm5* were grown in liquid MS medium for 10 days, and then were supplemented with 50µM ABA for 3h to induce expression of PP2CA. Next, ABA was washed, 50µM CHX was added and root samples were harvested at 0, 1, 3 and 6h. Total proteins were extracted by homogenizing the seedlings in lysis buffer (50mM Tris-HCl pH8.0, 150mM NaCl, 1% Triton X-100, 3mM DTT, 50µM proteasome inhibitor MG-132 and 1 tablet of protease inhibitor cocktail mini EDTA free Roche (1 tablet/10mL buffer)). The concentration of total protein was determined by Bradford assays and equal amounts of total proteins were mixed with 2× SDS loading buffer. Boiled samples were separated by SDS-PAGE gel electrophoresis and analyzed by immunoblot. Anti-PP2CA antibody was used to detect endogenous PP2CA. Actin was analyzed as a loading control using anti-Actin antibodies.

In vitro Ubiquitination Assay

For the *in vitro* substrate ubiquitination assay, a total reaction volume of 30µl was mixed by adding 500ng of purified MBP-RGLG1, 300ng of GST-PYL4 or GST-PYL8, 50ng of E1 (Sigma-Aldrich), 100ng of E2 UbcH5b (Enzo Life Sciences), 3µg FLAG-tagged ubiquitin (Sigma-Aldrich), 10mM phosphocreatine and 0.1U of creatine kinase in the ubiquitination buffer (50mM Tris-HCl pH 7.5, 3mM ATP, 5mM $MgCl_2$ and 0.5mM DTT), incubated at 37°C for 2 h. The reaction was stopped by adding 4xSDS loading buffer. Samples were separated by 8% sodium dodecyl sulfate polyacrylamide gel electrophoresis (SDS-PAGE) and analysed by Western blot.

In vivo Ubiquitination Assay of PP2CA-GFP in *N. benthamiana*

The synthetic Flag-Ub coding sequence was cloned in the pER8-GW vector to allow ß-estradiol inducible expression of tagged-Ub. *A. tumefaciens* cultures containing constructs that express constitutively PP2CA-GFP, HA-BPM3, the silencing suppressor p19 and ß-estradiol inducible Flag-Ub were co-infiltrated in *N. benthamiana* leaves. In control samples agrobacteria encoding PP2CA-GFP and the silencing suppressor p19 were co-infiltrated with empty agrobacteria. Two days after infiltration 1.5cm disk samples were collected, incubated for 16h in 100µM ß-estradiol, ground in liquid nitrogen and immediately placed in lysis buffer (50mM Tris-HCl pH8.0, 150mM NaCl, 1% Triton X-100, 3mM DTT, 50µM proteasome inhibitor MG-132 and 1 tablet of protease inhibitor cocktail mini EDTA free Roche (1 tablet/10mL buffer), 10nM ubiquitin aldehyde and 10nM N-ethylmaleimide (NEM)). GFP-tagged PP2CA was immunoprecipitated using super-paramagnetic micro MACS beads coupled to monoclonal anti-GFP antibody according to the manufacturer's instructions (Miltenyi Biotec). Purified immune-complexes were eluted in Laemmli buffer, boiled and run in a 10% SDS-PAGE gel. Proteins immunoprecipitated with anti-GFP antibody were transferred onto Immobilon-P membranes (Millipore) and probed with anti-GFP, anti-Flag, anti-Ub and anti-HA antibodies.

Affinity Purification of Ubiquitinated Proteins Using p62-agarose

Surface sterilized seeds of Col-0, *bpm3bpm5* and *amiR-bpm* plants were grown in liquid MS medium for 10 days and then were supplemented with 50µM ABA and 50µM MG-132 for 3h. Root plant material was collected, frozen in liquid nitrogen and extracted in 3 volumes of lysis buffer (50mM Tris-HCl pH8.0, 150mM NaCl, 1% Triton X-100, 3mM DTT, 50µM proteasome inhibitor MG-132 and 1 tablet of protease inhibitor cocktail mini EDTA free Roche (1 tablet/10mL buffer), 10nM ubiquitin aldehyde and 10nM NEM). After protein quantification of each plant extract, 300µg of total proteins were incubated with Ub-binding p62 agarose or with empty agarose. Pulled-down ubiquitinated proteins were eluted in Laemmli buffer, boiled and run in a 10% SDS-PAGE gel and analysed by immunoblotting. Detection of endogenous PP2CA was performed using anti-PP2CA and detection of ubiquitinated proteins using anti-Ub antibody.

In vivo Myristoylation Assay

RGLG1-GFP or RGLG1[G2A]-GFP were immunoprecipitated using super-paramagnetic micro MACS beads coupled to monoclonal anti-GFP antibody according to the manufacturer's instructions (Miltenyi Biotec). Immunoprecipitated proteins were analysed by SDS-PAGE and immunoblotted. To detect the myristoylation of RGLG1-GFP, the immunoprecipitated protein was probed with anti-myristic acid antibody followed by secondary anti-rat antibody. Detection was performed using the SuperSignal West Femto luminescence kit (Thermo Fisher Scientific).

Confocal Laser Scanning Microscopy

Confocal imaging was performed using a Zeiss LSM 780 AxioObserver.Z1 laser scanning microscope with C-Apochromat 40x/1.20-W corrective water immersion objective. The following fluorophores, which were excited and fluorescence emission detected by frame switching in the single or multi-tracking mode at the indicated wavelengths, were used in tobacco leaf infiltration experiments: GFP (488nm/500–530nm), OFP (561nm/575-600nm), SCFP (405nm/450-485nm), RFP (561nm/605-630nm) and YFP (488nm/529–550 nm). Pinholes were adjusted to one Air Unit for each wavelength. Post-acquisition image processing was performed using ZEN (ZEISS Efficient Navigation) Lite 2012 imaging software and NIH Image software ImageJ for image analyses.

Nuclei Staining and Counting

The reagent DAPI (Sigma-Aldrich) was used for staining nuclei. For the CLSM analysis hypocotyl/root cells of *A. thaliana* seedlings or agro-infiltrated *N. benthamiana* leaves were stained with 1ml of 5µg ml^{-1} DAPI solution and were analysed using a Zeiss LSM 780 AxioObserver.Z1 laser scanning microscope with C-Apochromat 403/1.20-W corrective water immersion objective. Leaf sections were mounted on a microscope slide and covered with distilled water for observation through the leaf abaxial side. Fluorescence was detected using the following excitation/emission parameters: 405nm/440–540nm and 488nm/500–530nm for DAPI and GFP channels, respectively. Post-acquisition image processing was performed using ZEN LITE 2012 imaging software and NIH Image software

ImageJ. Nuclei counting was performed in sections of
40.000µm² (n=20 fields) from five independent plants,
analysed through a full z-series of confocal images.

Mass Spectrometry Analysis

This protocol was performed by the Prof. Chengcai An's
laboratory. Pilot experiments to identify PP2CA-interacting
proteins using mass spectrometry analysis were performed
using transgenic A. thaliana lines expressing FLAG-PP2CA.
Two week-old seedlings were transferred from selective MS
medium to liquid MS containing 50µM MG-132 and 50µM
ABA for the indicated time. Plant material (300 mg) was
harvested, ground in liquid nitrogen and total proteins were
extracted in 3 volumes of lysis buffer (50mM Tris-HCl pH 7.5,
100mM NaCl, 0.1% NP-40, 50µM MG-132, 1mM DTT, 1mM
PMSF and plant-specific protease inhibitor cocktail). Total
protein concentration was determined by Bradford assays
(Bio-Rad), and each lysate (1ml) with equal amount of protein
was immunoprecipitated by incubating with 0.5µl anti-FLAG
and 25µl Protein G-Sepharose (Invitrogen) beads for 4 h at 4°C
with gentle rotation. After incubation, the beads were washed
with 1ml of lysis buffer for 4 times and eventually eluted by
adding 30µl 1× SDS protein loading buffer and boiling for
5min at 100°C. Immunoprecipitated proteins were separated
by 10% SDS-PAGE gel and PP2CA was detected using anti-
FLAG antibody.

The final experiments were performed with wo-week-old
transgenic *A. thaliana* seedlings overexpressing FLAG-
PP2CA treated with 50µM MG-132 50µM ABA for different
times. Approximately 4-5g each plant material was harvested
for total protein extraction. Anti-FLAG immunoprecipitates
were prepared as described above. FLAG-tagged proteins

were finally eluted from the beads by adding 400µg/ml 3×FLAG peptides in 1×PBS buffer with gentle rotation at 4°C for 1 h×3 times. The eluted proteins were then collected and lyophilized. A final volume of 30µl 1×PBS buffer was added to resuspend the powder. The resulting products were resolved on 4-12% NuPAGE® Bis-Tris MiNi Gel (Invitrogen) and visualized by colloidal blue staining kit (Invitrogen). The protein bands from 15 to 100Kd were cut from the gel and digested with trypsin. The extracted peptides were subjected to LC-MS/MS analysis using Easy nLC 1000 system (Thermo Scientific) connected to a Velos Pro Orbitrap Elite mass spectrometer (Thermo Scientific) equipped with a nano-ESI source. The raw data files were converted to mascot generic format (".mgf") using MSConvert before submitted for database search. Protein identification was carried using Mascot server v. 2.3.02 (Matrix Science) against TAIR Arabidopsis thaliana protein database

Statistics

Values are averages obtained from three independent experiments ±SDs. Significant differences were calculated using Student's T-test for single comparisons (* $p < 0.05$; ** $p < 0.01$) or ANOVA and Tukey post hoc test for multiple comparisons.

Fluorescence colocalization analysis was performed using the PSC colocalization plug-in of IMAGEJ (French et al., 2008). Pearson's (RP) and Spearman's (RS) correlation coefficients above 0.4 indicate colocalization, whereas lower values indicate a lack of colocalization.

REFERENCES

Abe H, Urao T, Ito T, Seki M, et al., 2003. Arabidopsis AtMYC2 (bHLH) and AtMYB2 (MYB) function as transcriptional activators in abscisic acid signaling. *Plant Cell* **15**(1), 63-78.

Akutsu M, Dikic I,Bremm A, 2016. Ubiquitin chain diversity at a glance. *J Cell Sci* **129**(5), 875-880.

Alam I, Cui D L, Batool K, Yang Y Q,Lu Y H, 2019. Comprehensive Genomic Survey, Characterization and Expression Analysis of the HECT Gene Family in Brassica rapa L. and Brassica oleracea L. *Genes (Basel)* **10**(5).

An F, Zhao Q, Ji Y, Li W, et al., 2010. Ethylene-induced stabilization of ETHYLENE INSENSITIVE3 and EIN3-LIKE1 is mediated by proteasomal degradation of EIN3 binding F-box 1 and 2 that requires EIN2 in Arabidopsis. *Plant Cell* **22**(7), 2384-2401.

Andersen P, Kragelund B B, Olsen A N, Larsen F H, et al., 2004. Structure and biochemical function of a prototypical Arabidopsis U-box domain. *J Biol Chem* **279**(38), 40053-40061.

Angers S, Li T, Yi X, MacCoss M J, et al., 2006. Molecular architecture and assembly of the DDB1-CUL4A ubiquitin ligase machinery. *Nature* **443**(7111), 590-593.

Antoni R, Gonzalez-Guzman M, Rodriguez L, Peirats-Llobet M, et al., 2013. PYRABACTIN RESISTANCE1-LIKE8 plays an important role for the regulation of abscisic acid signaling in root. *Plant Physiol* **161**(2), 931-941.

Antoni R, Gonzalez-Guzman M, Rodriguez L, Rodrigues A, et al., 2012. Selective inhibition of clade A phosphatases type 2C by PYR/PYL/RCAR abscisic acid receptors. *Plant Physiol* **158**(2), 970-980.

Arenas-Huertero F, Arroyo A, Zhou L, Sheen J,Leon P, 2000. Analysis of Arabidopsis glucose insensitive mutants, gin5 and gin6, reveals a central role of the plant hormone ABA in the regulation of plant vegetative development by sugar. *Genes Dev* **14**(16), 2085-2096.

Baek W, Lim C W, Luan S,Lee S C, 2019. The RING finger E3 ligases PIR1 and PIR2 mediate PP2CA degradation to enhance abscisic acid response in Arabidopsis. *Plant J.*

Bai C, Sen P, Hofmann K, Ma L, et al., 1996. SKP1 connects cell cycle regulators to the ubiquitin proteolysis machinery through a novel motif, the F-box. *Cell* **86**(2), 263-274.

Batistic O, Rehers M, Akerman A, Schlucking K, et al., 2012. S-acylation-dependent association of the calcium sensor CBL2 with the vacuolar membrane is essential for proper abscisic acid responses. *Cell Res* **22**(7), 1155-1168.

Belda-Palazon B, Gonzalez-Garcia M P, Lozano-Juste J, Coego A, et al., 2018. PYL8 mediates ABA perception in the root through non-cell-autonomous and ligand-stabilization-based mechanisms. *Proc Natl Acad Sci U S A* **115**(50), E11857-e11863.

Belda-Palazon B, Rodriguez L, Fernandez M A, Castillo M C, et al., 2016. FYVE1/FREE1 Interacts with the PYL4 ABA Receptor and Mediates its Delivery to the Vacuolar Degradation Pathway. *Plant Cell.*

Belda-Palazon B, Ruiz L, Marti E, Tarraga S, et al., 2012. Aminopropyltransferases involved in polyamine biosynthesis localize preferentially in the nucleus of plant cells. *PLoS One* **7**(10), e46907.

Bernhardt A, Lechner E, Hano P, Schade V, et al., 2006. CUL4 associates with DDB1 and DET1 and its downregulation affects diverse aspects of development in Arabidopsis thaliana. *Plant J* **47**(4), 591-603.

Berrocal-Lobo M, Stone S, Yang X, Antico J, et al., 2010. ATL9, a RING zinc finger protein with E3 ubiquitin ligase activity implicated in chitin- and NADPH oxidase-mediated defense responses. *PLoS One* **5**(12), e14426.

Bi C, Ma Y, Wang X F,Zhang D P, 2017. Overexpression of the transcription factor NF-YC9 confers abscisic acid hypersensitivity in Arabidopsis. *Plant Mol Biol* **95**(4-5), 425-439.

Bigeard J,Hirt H, 2018. Nuclear Signaling of Plant MAPKs. *Front Plant Sci* **9**, 469.

Boisson B, Giglione C,Meinnel T, 2003. Unexpected protein families including cell defense components feature in the N-myristoylome of a higher eukaryote. *J Biol Chem* **278**(44), 43418-43429.

Bosu D R,Kipreos E T, 2008. Cullin-RING ubiquitin ligases: global regulation and activation cycles. *Cell Div* **3**, 7.

Boudsocq M, Barbier-Brygoo H,Lauriere C, 2004. Identification of nine sucrose nonfermenting 1-related protein kinases 2 activated by hyperosmotic and saline stresses in Arabidopsis thaliana. *J Biol Chem* **279**(40), 41758-41766.

Brandt B, Munemasa S, Wang C, Nguyen D, et al., 2015. Calcium specificity signaling mechanisms in abscisic acid signal transduction in Arabidopsis guard cells. *Elife* **4**.

Bremm A,Komander D, 2011. Emerging roles for Lys11-linked polyubiquitin in cellular regulation. *Trends Biochem Sci* **36**(7), 355-363.

Bueso E, Ibanez C, Sayas E, Munoz-Bertomeu J, et al., 2014a. A forward genetic approach in Arabidopsis thaliana identifies a RING-type ubiquitin ligase as a novel determinant of seed longevity. *Plant Sci* **215-216**, 110-116.

Bueso E, Rodriguez L, Lorenzo-Orts L, Gonzalez-Guzman M, et al., 2014b. The single-subunit RING-type E3 ubiquitin ligase RSL1 targets PYL4 and PYR1 ABA receptors in plasma membrane to modulate abscisic acid signaling. *Plant J* **80**(6), 1057-1071.

Burnaevskiy N, Fox T G, Plymire D A, Ertelt J M, et al., 2013. Proteolytic elimination of N-myristoyl modifications by the Shigella virulence factor IpaJ. *Nature* **496**(7443), 106-109.

Burnaevskiy N, Peng T, Reddick L E, Hang H C,Alto N M, 2015. Myristoylome profiling reveals a concerted mechanism of ARF GTPase deacylation by the bacterial protease IpaJ. *Mol Cell* **58**(1), 110-122.

Cai Z, Liu J, Wang H, Yang C, et al., 2014. GSK3-like kinases positively modulate abscisic acid signaling through phosphorylating subgroup III SnRK2s in Arabidopsis. *Proc Natl Acad Sci U S A* **111**(26), 9651-9656.

Callis J, 2014. The ubiquitination machinery of the ubiquitin system. *Arabidopsis Book* **12**, e0174.

Callis J,Vierstra R, 1989. Ubiquitin and ubiquitin genes in higher plants. *Oxford surveys of plant molecular and cell biology.*

Cardozo T,Pagano M, 2004. The SCF ubiquitin ligase: insights into a molecular machine. *Nat Rev Mol Cell Biol* **5**(9), 739-751.

Castillo M C, Lozano-Juste J, Gonzalez-Guzman M, Rodriguez L, et al., 2015. Inactivation of PYR/PYL/RCAR ABA receptors by tyrosine nitration may enable rapid inhibition of ABA signaling by nitric oxide in plants. *Sci Signal* **8**(392), ra89.

Chatr-Aryamontri A, van der Sloot A,Tyers M, 2018. At Long Last, a C-Terminal Bookend for the Ubiquitin Code. *Mol Cell* **70**(4), 568-571.

Chaumet A, Wright G D, Seet S H, Tham K M, et al., 2015. Nuclear envelope-associated endosomes deliver surface proteins to the nucleus. *Nat Commun* **6**, 8218.

Chen H H, Qu L, Xu Z H, Zhu J K,Xue H W, 2018. EL1-like Casein Kinases Suppress ABA Signaling and Responses by Phosphorylating and Destabilizing the ABA Receptors PYR/PYLs in Arabidopsis. *Mol Plant* **11**(5), 706-719.

Chen J, Yu F, Liu Y, Du C, et al., 2016. FERONIA interacts with ABI2-type phosphatases to facilitate signaling cross-talk between abscisic acid and RALF peptide in Arabidopsis. *Proc Natl Acad Sci U S A* **113**(37), E5519-5527.

Chen L, Bernhardt A, Lee J,Hellmann H, 2015. Identification of Arabidopsis MYB56 as a novel substrate for CRL3(BPM) E3 ligases. *Mol Plant* **8**(2), 242-250.

Chen L, Lee J H, Weber H, Tohge T, et al., 2013. Arabidopsis BPM proteins function as substrate adaptors to a cullin3-based E3 ligase to affect fatty acid metabolism in plants. *Plant Cell* **25**(6), 2253-2264.

Cheng A, Kaldis P,Solomon M J, 2000. Dephosphorylation of human cyclin-dependent kinases by protein phosphatase type 2C alpha and beta 2 isoforms. *J Biol Chem* **275**(44), 34744-34749.

Cheng C, Wang Z, Ren Z, Zhi L, et al., 2017. SCFAtPP2-B11 modulates ABA signaling by facilitating SnRK2.3 degradation in Arabidopsis thaliana. *PLoS Genet* **13**(8), e1006947.

Cheng M C, Hsieh E J, Chen J H, Chen H Y,Lin T P, 2012. Arabidopsis RGLG2, functioning as a RING E3 ligase, interacts with AtERF53 and negatively regulates the plant drought stress response. *Plant Physiol* **158**(1), 363-375.

Cherel I, Michard E, Platet N, Mouline K, et al., 2002. Physical and functional interaction of the Arabidopsis K(+) channel AKT2 and phosphatase AtPP2CA. *Plant Cell* **14**(5), 1133-1146.

Chini A, Fonseca S, Fernandez G, Adie B, et al., 2007. The JAZ family of repressors is the missing link in jasmonate signalling. *Nature* **448**(7154), 666-671.

Choi H, Hong J, Ha J, Kang J,Kim S Y, 2000. ABFs, a family of ABA-responsive element binding factors. *J Biol Chem* **275**(3), 1723-1730.

Ciechanover A, Heller H, Elias S, Haas A L,Hershko A, 1980. ATP-dependent conjugation of reticulocyte proteins with the polypeptide required for protein degradation. *Proc Natl Acad Sci U S A* **77**(3), 1365-1368.

Criqui M C, de Almeida Engler J, Camasses A, Capron A, et al., 2002. Molecular characterization of plant ubiquitin-conjugating enzymes belonging to the UbcP4/E2-C/UBCx/UbcH10 gene family. *Plant Physiol* **130**(3), 1230-1240.

Cutler S R, Rodriguez P L, Finkelstein R R,Abrams S R, 2010. Abscisic acid: emergence of a core signaling network. *Annu Rev Plant Biol* **61**, 651-679.

de Castro E, Sigrist C J, Gattiker A, Bulliard V, et al., 2006. ScanProsite: detection of PROSITE signature matches and ProRule-associated functional and structural residues in proteins. *Nucleic Acids Res* **34**(Web Server issue), W362-365.

Deblaere R, Bytebier B, De Greve H, Deboeck F, et al., 1985. Efficient octopine Ti plasmid-derived vectors for Agrobacterium-mediated gene transfer to plants. *Nucleic Acids Res* **13**(13), 4777-4788.

Del Pozo J C,Manzano C, 2014. Auxin and the ubiquitin pathway. Two players-one target: the cell cycle in action. *J Exp Bot* **65**(10), 2617-2632.

Dharmasiri N, Dharmasiri S, Weijers D, Lechner E, et al., 2005. Plant development is regulated by a family of auxin receptor F box proteins. *Dev Cell* **9**(1), 109-119.

Dietrich D, Pang L, Kobayashi A, Fozard J A, et al., 2017. Root hydrotropism is controlled via a cortex-specific growth mechanism. *Nat Plants* **3**, 17057.

Ding Y, Lv J, Shi Y, Gao J, et al., 2019. EGR2 phosphatase regulates OST1 kinase activity and freezing tolerance in Arabidopsis. *Embo j* **38**(1).

Ding Y, Sun T, Ao K, Peng Y, et al., 2018. Opposite Roles of Salicylic Acid Receptors NPR1 and NPR3/NPR4 in Transcriptional Regulation of Plant Immunity. *Cell* **173**(6), 1454-1467.e1415.

Dissmeyer N, 2019. Conditional Protein Function via N-Degron Pathway-Mediated Proteostasis in Stress Physiology. *Annu Rev Plant Biol* **70**, 83-117.

Dittrich M, Mueller H M, Bauer H, Peirats-Llobet M, et al., 2019. The role of Arabidopsis ABA receptors from the PYR/PYL/RCAR family in stomatal acclimation and closure signal integration. *Nat Plants*.

Dove K K, Stieglitz B, Duncan E D, Rittinger K,Klevit R E, 2016. Molecular insights into RBR E3 ligase ubiquitin transfer mechanisms. *EMBO Rep* **17**(8), 1221-1235.

Downes B P, Stupar R M, Gingerich D J,Vierstra R D, 2003. The HECT ubiquitin-protein ligase (UPL) family in Arabidopsis: UPL3 has a specific role in trichome development. *Plant J* **35**(6), 729-742.

Duda D M, Borg L A, Scott D C, Hunt H W, et al., 2008. Structural insights into NEDD8 activation of cullin-RING ligases: conformational control of conjugation. *Cell* **134**(6), 995-1006.

Duda D M, Olszewski J L, Schuermann J P, Kurinov I, et al., 2013. Structure of HHARI, a RING-IBR-RING ubiquitin ligase: autoinhibition of an Ariadne-family E3 and insights into ligation mechanism. *Structure* **21**(6), 1030-1041.

Duek P D,Fankhauser C, 2005. bHLH class transcription factors take centre stage in phytochrome signalling. *Trends Plant Sci* **10**(2), 51-54.

Dupeux F, Antoni R, Betz K, Santiago J, et al., 2011a. Modulation of abscisic acid signaling in vivo by an engineered receptor-insensitive protein phosphatase type 2C allele. *Plant Physiol* **156**(1), 106-116.

Dupeux F, Santiago J, Betz K, Twycross J, et al., 2011b. A thermodynamic switch modulates abscisic acid receptor sensitivity. *Embo j* **30**(20), 4171-4184.

Dupre S,Haguenauer-Tsapis R, 2001. Deubiquitination step in the endocytic pathway of yeast plasma membrane proteins: crucial role of Doa4p ubiquitin isopeptidase. *Mol Cell Biol* **21**(14), 4482-4494.

Eddins M J, Varadan R, Fushman D, Pickart C M,Wolberger C, 2007. Crystal structure and solution NMR studies of Lys48-linked tetraubiquitin at neutral pH. *J Mol Biol* **367**(1), 204-211.

Edel K H,Kudla J, 2016. Integration of calcium and ABA signaling. *Curr Opin Plant Biol* **33**, 83-91.

Ezcurra I, Wycliffe P, Nehlin L, Ellerstrom M,Rask L, 2000. Transactivation of the Brassica napus napin promoter by ABI3 requires interaction of the conserved B2 and B3 domains of ABI3 with different cis-elements: B2 mediates activation through an ABRE, whereas B3 interacts with an RY/G-box. *Plant J* **24**(1), 57-66.

Feldman R M, Correll C C, Kaplan K B,Deshaies R J, 1997. A complex of Cdc4p, Skp1p, and Cdc53p/cullin catalyzes ubiquitination of the phosphorylated CDK inhibitor Sic1p. *Cell* **91**(2), 221-230.

Feng J,Shen W H, 2014. Dynamic regulation and function of histone monoubiquitination in plants. *Front Plant Sci* **5**, 83.

Finkelstein R, Lynch T, Reeves W, Petitfils M,Mostachetti M, 2011. Accumulation of the transcription factor ABA-insensitive (ABI)4 is tightly regulated post-transcriptionally. *J Exp Bot* **62**(11), 3971-3979.

Finkelstein R R, Wang M L, Lynch T J, Rao S,Goodman H M, 1998. The Arabidopsis abscisic acid response locus ABI4 encodes an APETALA 2 domain protein. *Plant Cell* **10**(6), 1043-1054.

Forster S, Schmidt L K, Kopic E, Anschutz U, et al., 2019. Wounding-Induced Stomatal Closure Requires Jasmonate-Mediated Activation of GORK K(+) Channels by a Ca(2+) Sensor-Kinase CBL1-CIPK5 Complex. *Dev Cell* **48**(1), 87-99.e86.

French A P, Mills S, Swarup R, Bennett M J,Pridmore T P, 2008. Colocalization of fluorescent markers in confocal microscope images of plant cells. *Nat Protoc* **3**(4), 619-628.

Fu Z Q, Yan S, Saleh A, Wang W, et al., 2012. NPR3 and NPR4 are receptors for the immune signal salicylic acid in plants. *Nature* **486**(7402), 228-232.

Fujii H, Chinnusamy V, Rodrigues A, Rubio S, et al., 2009. In vitro reconstitution of an abscisic acid signalling pathway. *Nature* **462**(7273), 660-664.

Fujii H, Verslues P E,Zhu J K, 2007. Identification of two protein kinases required for abscisic acid regulation of seed germination, root growth, and gene expression in Arabidopsis. *Plant Cell* **19**(2), 485-494.

Fujii H, Verslues P E,Zhu J K, 2011. Arabidopsis decuple mutant reveals the importance of SnRK2 kinases in osmotic stress responses in vivo. *Proc Natl Acad Sci U S A* **108**(4), 1717-1722.

Fujii H,Zhu J K, 2009. Arabidopsis mutant deficient in 3 abscisic acid-activated protein kinases reveals critical roles in growth, reproduction, and stress. *Proc Natl Acad Sci U S A* **106**(20), 8380-8385.

Fujita Y, Fujita M, Satoh R, Maruyama K, et al., 2005. AREB1 is a transcription activator of novel ABRE-dependent ABA signaling that enhances drought stress tolerance in Arabidopsis. *Plant Cell* **17**(12), 3470-3488.

Fujita Y, Nakashima K, Yoshida T, Katagiri T, et al., 2009. Three SnRK2 protein kinases are the main positive regulators of abscisic acid signaling in response to water stress in Arabidopsis. *Plant Cell Physiol* **50**(12), 2123-2132.

Fujita Y, Yoshida T,Yamaguchi-Shinozaki K, 2013. Pivotal role of the AREB/ABF-SnRK2 pathway in ABRE-mediated transcription in response to osmotic stress in plants. *Physiol Plant* **147**(1), 15-27.

Fulop K, Tarayre S, Kelemen Z, Horvath G, et al., 2005. Arabidopsis anaphase-promoting complexes: multiple activators and wide range of substrates might keep APC perpetually busy. *Cell Cycle* **4**(8), 1084-1092.

Furniss J J, Grey H, Wang Z, Nomoto M, et al., 2018. Proteasome-associated HECT-type ubiquitin ligase activity is required for plant immunity. *PLoS Pathog* **14**(11), e1007447.

Gagne J M, Smalle J, Gingerich D J, Walker J M, et al., 2004. Arabidopsis EIN3-binding F-box 1 and 2 form ubiquitin-protein ligases that repress ethylene action and promote growth by directing EIN3 degradation. *Proc Natl Acad Sci U S A* **101**(17), 6803-6808.

Galan J M,Haguenauer-Tsapis R, 1997. Ubiquitin lys63 is involved in ubiquitination of a yeast plasma membrane protein. *Embo j* **16**(19), 5847-5854.

Gao C, Luo M, Zhao Q, Yang R, et al., 2014. A unique plant ESCRT component, FREE1, regulates multivesicular body protein sorting and plant growth. *Curr Biol* **24**(21), 2556-2563.

Gao C, Zhuang X, Cui Y, Fu X, et al., 2015. Dual roles of an Arabidopsis ESCRT component FREE1 in regulating vacuolar protein transport and autophagic degradation. *Proc Natl Acad Sci U S A* **112**(6), 1886-1891.

Garcia-Leon M, Cuyas L, Abd El-Moneim D, Rodriguez L, et al., 2019. Stomatal aperture and turnover of ABA receptors are regulated by Arabidopsis ALIX. *Plant Cell.*

Gehl C, Waadt R, Kudla J, Mendel R R,Hansch R, 2009. New GATEWAY vectors for high throughput analyses of protein-protein interactions by bimolecular fluorescence complementation. *Mol Plant* **2**(5), 1051-1058.

Geiger D, Scherzer S, Mumm P, Stange A, et al., 2009. Activity of guard cell anion channel SLAC1 is controlled by drought-stress signaling kinase-phosphatase pair. *Proc Natl Acad Sci U S A* **106**(50), 21425-21430.

Gingerich D J, Gagne J M, Salter D W, Hellmann H, et al., 2005. Cullins 3a and 3b assemble with members of the broad complex/tramtrack/bric-a-brac (BTB) protein family to form essential ubiquitin-protein ligases (E3s) in Arabidopsis. *J Biol Chem* **280**(19), 18810-18821.

Gingerich D J, Hanada K, Shiu S H,Vierstra R D, 2007. Large-scale, lineage-specific expansion of a bric-a-brac/tramtrack/broad complex ubiquitin-ligase gene family in rice. *Plant Cell* **19**(8), 2329-2348.

Giraud E, Van Aken O, Ho L H,Whelan J, 2009. The transcription factor ABI4 is a regulator of mitochondrial retrograde expression of ALTERNATIVE OXIDASE1a. *Plant Physiol* **150**(3), 1286-1296.

Goda H, Sasaki E, Akiyama K, Maruyama-Nakashita A, et al., 2008. The AtGenExpress hormone and chemical treatment data set: experimental design, data evaluation, model data analysis and data access. *Plant J* **55**(3), 526-542.

Gonzalez-Grandio E, Pajoro A, Franco-Zorrilla J M, Tarancon C, et al., 2017. Abscisic acid signaling is controlled by a BRANCHED1/HD-ZIP I cascade in Arabidopsis axillary buds. *Proc Natl Acad Sci U S A* **114**(2), E245-e254.

Gonzalez-Guzman M, Pizzio G A, Antoni R, Vera-Sirera F, et al., 2012. Arabidopsis PYR/PYL/RCAR receptors play a major role in quantitative regulation of stomatal aperture and transcriptional response to abscisic acid. *Plant Cell* **24**(6), 2483-2496.

Gonzalez-Guzman M, Rodriguez L, Lorenzo-Orts L, Pons C, et al., 2014. Tomato PYR/PYL/RCAR abscisic acid receptors show high expression in root, differential sensitivity to the abscisic acid agonist quinabactin, and the capability to enhance plant drought resistance. *J Exp Bot* **65**(15), 4451-4464.

Gosti F, Beaudoin N, Serizet C, Webb A A, et al., 1999. ABI1 protein phosphatase 2C is a negative regulator of abscisic acid signaling. *Plant Cell* **11**(10), 1897-1910.

Graciet E, Mesiti F,Wellmer F, 2010. Structure and evolutionary conservation of the plant N-end rule pathway. *Plant J* **61**(5), 741-751.

Gray W M, del Pozo J C, Walker L, Hobbie L, et al., 1999. Identification of an SCF ubiquitin-ligase complex required for auxin response in Arabidopsis thaliana. *Genes Dev* **13**(13), 1678-1691.

Gray W M, Kepinski S, Rouse D, Leyser O,Estelle M, 2001. Auxin regulates SCFTIR1-dependent degradation of AUX/IAA proteins. *Nature* **414**(6861), 271-276.

Grondin A, Rodrigues O, Verdoucq L, Merlot S, et al., 2015. Aquaporins Contribute to ABA-Triggered Stomatal Closure through OST1-Mediated Phosphorylation. *Plant Cell* **27**(7), 1945-1954.

Guo Y, Xiong L, Song C P, Gong D, et al., 2002. A calcium sensor and its interacting protein kinase are global regulators of abscisic acid signaling in Arabidopsis. *Dev Cell* **3**(2), 233-244.

Haglund K, Di Fiore P P,Dikic I, 2003. Distinct monoubiquitin signals in receptor endocytosis. *Trends Biochem Sci* **28**(11), 598-603.

He L, Bian J, Xu J,Yang K, 2019. Novel Maize NAC Transcriptional Repressor ZmNAC071 Confers Enhanced Sensitivity to ABA and Osmotic Stress by Downregulating Stress-Responsive Genes in Transgenic Arabidopsis. *J Agric Food Chem* **67**(32), 8905-8918.

He Y, Hao Q, Li W, Yan C, et al., 2014. Identification and characterization of ABA receptors in Oryza sativa. *PLoS One* **9**(4), e95246.

He Z, Zhong J, Sun X, Wang B, et al., 2018. The Maize ABA Receptors ZmPYL8, 9, and 12 Facilitate Plant Drought Resistance. *Front Plant Sci* **9**, 422.

Herranz M C, Pallas V,Aparicio F, 2012. Multifunctional roles for the N-terminal basic motif of Alfalfa mosaic virus coat protein: nucleolar/cytoplasmic shuttling, modulation of RNA-binding activity, and virion formation. *Mol Plant Microbe Interact* **25**(8), 1093-1103.

Himmelbach A, Hoffmann T, Leube M, Hohener B,Grill E, 2002. Homeodomain protein ATHB6 is a target of the protein phosphatase ABI1 and regulates hormone responses in Arabidopsis. *Embo j* **21**(12), 3029-3038.

Hornacek M, Kovacik L, Mazel T, Cmarko D, et al., 2017. Fluctuations of pol I and fibrillarin contents of the nucleoli. *Nucleus* 8(4), 421-432.

Hoth S, Morgante M, Sanchez J P, Hanafey M K, et al., 2002. Genome-wide gene expression profiling in Arabidopsis thaliana reveals new targets of abscisic acid and largely impaired gene regulation in the abi1-1 mutant. *J Cell Sci* 115(Pt 24), 4891-4900.

Hrabak E M, Chan C W, Gribskov M, Harper J F, et al., 2003. The Arabidopsis CDPK-SnRK superfamily of protein kinases. *Plant Physiol* 132(2), 666-680.

Hua Z,Kao T H, 2006. Identification and characterization of components of a putative petunia S-locus F-box-containing E3 ligase complex involved in S-RNase-based self-incompatibility. *Plant Cell* 18(10), 2531-2553.

Hua Z,Vierstra R D, 2011. The cullin-RING ubiquitin-protein ligases. *Annu Rev Plant Biol* 62, 299-334.

Hurley J H,Ren X, 2009. The circuitry of cargo flux in the ESCRT pathway. *J Cell Biol* 185(2), 185-187.

Imes D, Mumm P, Bohm J, Al-Rasheid K A, et al., 2013. Open stomata 1 (OST1) kinase controls R-type anion channel QUAC1 in Arabidopsis guard cells. *Plant J* 74(3), 372-382.

Irigoyen M L, Iniesto E, Rodriguez L, Puga M I, et al., 2014. Targeted degradation of abscisic acid receptors is mediated by the ubiquitin ligase substrate adaptor DDA1 in Arabidopsis. *Plant Cell* 26(2), 712-728.

Jakobson L, Lindgren L O, Verdier G, Laanemets K, et al., 2016. BODYGUARD is required for the biosynthesis of cutin in Arabidopsis. *New Phytol* 211(2), 614-626.

Kerchev P I, Pellny T K, Vivancos P D, Kiddle G, et al., 2011. The transcription factor ABI4 Is required for the ascorbic acid-dependent regulation of growth and regulation of jasmonate-dependent defense signaling pathways in Arabidopsis. *Plant Cell* 23(9), 3319-3334.

Kim H C,Huibregtse J M, 2009. Polyubiquitination by HECT E3s and the determinants of chain type specificity. *Mol Cell Biol* 29(12), 3307-3318.

Kim S J,Kim W T, 2013. Suppression of Arabidopsis RING E3 ubiquitin ligase AtATL78 increases tolerance to cold stress and decreases tolerance to drought stress. *FEBS Lett* **587**(16), 2584-2590.

Kim T H, Bohmer M, Hu H, Nishimura N,Schroeder J I, 2010. Guard cell signal transduction network: advances in understanding abscisic acid, CO2, and Ca2+ signaling. *Annu Rev Plant Biol* **61**, 561-591.

Kiyosue T, Yamaguchi-Shinozaki K,Shinozaki K, 1994. Cloning of cDNAs for genes that are early-responsive to dehydration stress (ERDs) in Arabidopsis thaliana L.: identification of three ERDs as HSP cognate genes. *Plant Mol Biol* **25**(5), 791-798.

Kleiger G, Saha A, Lewis S, Kuhlman B,Deshaies R J, 2009. Rapid E2-E3 assembly and disassembly enable processive ubiquitylation of cullin-RING ubiquitin ligase substrates. *Cell* **139**(5), 957-968.

Kolb C, Nagel M K, Kalinowska K, Hagmann J, et al., 2015. FYVE1 is essential for vacuole biogenesis and intracellular trafficking in Arabidopsis. *Plant Physiol* **167**(4), 1361-1373.

Kollist T, Moldau H, Rasulov B, Oja V, et al., 2007. A novel device detects a rapid ozone-induced transient stomatal closure in intact Arabidopsis and its absence in abi2 mutant. *Physiol Plant* **129**(4), 796-803.

Komander D,Rape M, 2012. The ubiquitin code. *Annu Rev Biochem* **81**, 203-229.

Kong L, Cheng J, Zhu Y, Ding Y, et al., 2015. Degradation of the ABA co-receptor ABI1 by PUB12/13 U-box E3 ligases. *Nat Commun* **6**, 8630.

Konishi M,Yanagisawa S, 2008. Ethylene signaling in Arabidopsis involves feedback regulation via the elaborate control of EBF2 expression by EIN3. *Plant J* **55**(5), 821-831.

Koren I, Timms R T, Kula T, Xu Q, et al., 2018. The Eukaryotic Proteome Is Shaped by E3 Ubiquitin Ligases Targeting C-Terminal Degrons. *Cell* **173**(7), 1622-1635.e1614.

Kosugi S, Hasebe M, Tomita M,Yanagawa H, 2009. Systematic identification of cell cycle-dependent yeast nucleocytoplasmic shuttling proteins by prediction of composite motifs. *Proc Natl Acad Sci U S A* **106**(25), 10171-10176.

Koussevitzky S, Nott A, Mockler T C, Hong F, et al., 2007. Signals from chloroplasts converge to regulate nuclear gene expression. *Science* **316**(5825), 715-719.

Kowarschik K, Hoehenwarter W, Marillonnet S,Trujillo M, 2018. UbiGate: a synthetic biology toolbox to analyse ubiquitination. *New Phytol* **217**(4), 1749-1763.

Kraft E, Stone S L, Ma L, Su N, et al., 2005. Genome analysis and functional characterization of the E2 and RING-type E3 ligase ubiquitination enzymes of Arabidopsis. *Plant Physiol* **139**(4), 1597-1611.

Kuhn J M, Boisson-Dernier A, Dizon M B, Maktabi M H,Schroeder J I, 2006. The protein phosphatase AtPP2CA negatively regulates abscisic acid signal transduction in Arabidopsis, and effects of abh1 on AtPP2CA mRNA. *Plant Physiol* **140**(1), 127-139.

Kumar D, Kumar R, Baek D, Hyun T K, et al., 2017. Arabidopsis thaliana RECEPTOR DEAD KINASE1 Functions as a Positive Regulator in Plant Responses to ABA. *Mol Plant* **10**(2), 223-243.

Kushwaha H R, Joshi R, Pareek A,Singla-Pareek S L, 2016. MATH-Domain Family Shows Response toward Abiotic Stress in Arabidopsis and Rice. *Front Plant Sci* **7**, 923.

Lang-Mladek C, Xie L, Nigam N, Chumak N, et al., 2012. UV-B signaling pathways and fluence rate dependent transcriptional regulation of ARIADNE12. *Physiol Plant* **145**(4), 527-539.

Lange O F, Lakomek N A, Fares C, Schroder G F, et al., 2008. Recognition dynamics up to microseconds revealed from an RDC-derived ubiquitin ensemble in solution. *Science* **320**(5882), 1471-1475.

Lau O S,Deng X W, 2012. The photomorphogenic repressors COP1 and DET1: 20 years later. *Trends Plant Sci* **17**(10), 584-593.

Lechner E, Leonhardt N, Eisler H, Parmentier Y, et al., 2011. MATH/BTB CRL3 receptors target the homeodomain-leucine zipper ATHB6 to modulate abscisic acid signaling. *Dev Cell* **21**(6), 1116-1128.

Lechner E, Xie D, Grava S, Pigaglio E, et al., 2002. The AtRbx1 protein is part of plant SCF complexes, and its down-regulation causes severe growth and developmental defects. *J Biol Chem* **277**(51), 50069-50080.

Lee H J, Park Y J, Seo P J, Kim J H, et al., 2015. Systemic Immunity Requires SnRK2.8-Mediated Nuclear Import of NPR1 in Arabidopsis. *Plant Cell* **27**(12), 3425-3438.

Lee J H, Terzaghi W, Gusmaroli G, Charron J B, et al., 2008. Characterization of Arabidopsis and rice DWD proteins and their roles as substrate receptors for CUL4-RING E3 ubiquitin ligases. *Plant Cell* **20**(1), 152-167.

Lee S C, Lan W, Buchanan B B,Luan S, 2009. A protein kinase-phosphatase pair interacts with an ion channel to regulate ABA signaling in plant guard cells. *Proc Natl Acad Sci U S A* **106**(50), 21419-21424.

Leitner J, Petrasek J, Tomanov K, Retzer K, et al., 2012. Lysine63-linked ubiquitylation of PIN2 auxin carrier protein governs hormonally controlled adaptation of Arabidopsis root growth. *Proc Natl Acad Sci U S A* **109**(21), 8322-8327.

Leung J, Bouvier-Durand M, Morris P C, Guerrier D, et al., 1994. Arabidopsis ABA response gene ABI1: features of a calcium-modulated protein phosphatase. *Science* **264**(5164), 1448-1452.

Leung J, Merlot S,Giraudat J, 1997. The Arabidopsis ABSCISIC ACID-INSENSITIVE2 (ABI2) and ABI1 genes encode homologous protein phosphatases 2C involved in abscisic acid signal transduction. *Plant Cell* **9**(5), 759-771.

Li D, Zhang L, Li X, Kong X, et al., 2018. AtRAE1 is involved in degradation of ABA receptor RCAR1 and negatively regulates ABA signalling in Arabidopsis. *Plant Cell Environ* **41**(1), 231-244.

Li W,Schmidt W, 2010. A lysine-63-linked ubiquitin chain-forming conjugase, UBC13, promotes the developmental responses to iron deficiency in Arabidopsis roots. *Plant J* **62**(2), 330-343.

Li W, Wang L, Sheng X, Yan C, et al., 2013. Molecular basis for the selective and ABA-independent inhibition of PP2CA by PYL13. *Cell Res* **23**(12), 1369-1379.

Li X, Chang Y, Ma S, Shen J, et al., 2019a. Genome-Wide Identification of SNAC1-Targeted Genes Involved in Drought Response in Rice. *Front Plant Sci* **10**, 982.

Li X, Kong X, Huang Q, Zhang Q, et al., 2019b. CARK1 phosphorylates subfamily III members of ABA receptors. *J Exp Bot* **70**(2), 519-528.

Li Y, Zhang L, Li D, Liu Z, et al., 2016. The Arabidopsis F-box E3 ligase RIFP1 plays a negative role in abscisic acid signalling by facilitating ABA receptor RCAR3 degradation. *Plant Cell Environ* **39**(3), 571-582.

Lin H C, Yeh C W, Chen Y F, Lee T T, et al., 2018. C-Terminal End-Directed Protein Elimination by CRL2 Ubiquitin Ligases. *Mol Cell* **70**(4), 602-613.e603.

Lin Q, Wu F, Sheng P, Zhang Z, et al., 2015. The SnRK2-APC/C(TE) regulatory module mediates the antagonistic action of gibberellic acid and abscisic acid pathways. *Nat Commun* **6**, 7981.

Ling Q, Huang W, Baldwin A,Jarvis P, 2012. Chloroplast biogenesis is regulated by direct action of the ubiquitin-proteasome system. *Science* **338**(6107), 655-659.

Liu Y C, Wu Y R, Huang X H, Sun J,Xie Q, 2011. AtPUB19, a U-box E3 ubiquitin ligase, negatively regulates abscisic acid and drought responses in Arabidopsis thaliana. *Mol Plant* **4**(6), 938-946.

Livak K J,Schmittgen T D, 2001. Analysis of relative gene expression data using real-time quantitative PCR and the 2(-Delta Delta C(T)) Method. *Methods* **25**(4), 402-408.

Lopez-Molina L,Chua N H, 2000. A null mutation in a bZIP factor confers ABA-insensitivity in Arabidopsis thaliana. *Plant Cell Physiol* **41**(5), 541-547.

Lopez-Molina L, Mongrand S,Chua N H, 2001. A postgermination developmental arrest checkpoint is mediated by abscisic acid and requires the ABI5 transcription factor in Arabidopsis. *Proc Natl Acad Sci U S A* **98**(8), 4782-4787.

Lopez-Molina L, Mongrand S, McLachlin D T, Chait B T,Chua N H, 2002. ABI5 acts downstream of ABI3 to execute an ABA-dependent growth arrest during germination. *Plant J* **32**(3), 317-328.

Lozano-Juste J,Leon J, 2010. Nitric oxide modulates sensitivity to ABA. *Plant Signal Behav* **5**(3), 314-316.

Lu D, Lin W, Gao X, Wu S, et al., 2011. Direct ubiquitination of pattern recognition receptor FLS2 attenuates plant innate immunity. *Science* **332**(6036), 1439-1442.

Luan S, 2003. Protein phosphatases in plants. *Annu Rev Plant Biol* **54**, 63-92.

Lyapina S, Cope G, Shevchenko A, Serino G, et al., 2001. Promotion of NEDD-CUL1 conjugate cleavage by COP9 signalosome. *Science* **292**(5520), 1382-1385.

Lynch T, Erickson B J,Finkelstein R R, 2012. Direct interactions of ABA-insensitive(ABI)-clade protein phosphatase(PP)2Cs with calcium-dependent protein kinases and ABA response element-binding bZIPs may contribute to turning off ABA response. *Plant Mol Biol* **80**(6), 647-658.

Ma Y, Szostkiewicz I, Korte A, Moes D, et al., 2009. Regulators of PP2C phosphatase activity function as abscisic acid sensors. *Science* **324**(5930), 1064-1068.

Majeran W, Le Caer J P, Ponnala L, Meinnel T,Giglione C, 2018. Targeted Profiling of Arabidopsis thaliana Subproteomes Illuminates Co- and Posttranslationally N-Terminal Myristoylated Proteins. *Plant Cell* **30**(3), 543-562.

March E,Farrona S, 2017. Plant Deubiquitinases and Their Role in the Control of Gene Expression Through Modification of Histones. *Front Plant Sci* **8**, 2274.

Marin I, 2010. Diversification and Specialization of Plant RBR Ubiquitin Ligases. *PLoS One* **5**(7), e11579.

McCullough J, Clague M J,Urbe S, 2004. AMSH is an endosome-associated ubiquitin isopeptidase. *J Cell Biol* **166**(4), 487-492.

Melcher K, Ng L M, Zhou X E, Soon F F, et al., 2009. A gate-latch-lock mechanism for hormone signalling by abscisic acid receptors. *Nature* **462**(7273), 602-608.

Merchante C, Alonso J M,Stepanova A N, 2013. Ethylene signaling: simple ligand, complex regulation. *Curr Opin Plant Biol* **16**(5), 554-560.

Meyer H J,Rape M, 2014. Enhanced protein degradation by branched ubiquitin chains. *Cell* **157**(4), 910-921.

Meyer K, Leube M P,Grill E, 1994. A protein phosphatase 2C involved in ABA signal transduction in Arabidopsis thaliana. *Science* **264**(5164), 1452-1455.

Miao C, Xiao L, Hua K, Zou C, et al., 2018. Mutations in a subfamily of abscisic acid receptor genes promote rice growth and productivity. *Proc Natl Acad Sci U S A* **115**(23), 6058-6063.

Milborrow B V, 1974. The Chemistry and Physiology of Abscisic Acid. *Annual Review of Plant Physiology* **25**(1), 259-307.

Min K W, Hwang J W, Lee J S, Park Y, et al., 2003. TIP120A associates with cullins and modulates ubiquitin ligase activity. *J Biol Chem* **278**(18), 15905-15910.

Mitula F, Tajdel M, Ciesla A, Kasprowicz-Maluski A, et al., 2015. Arabidopsis ABA-Activated Kinase MAPKKK18 is Regulated by Protein Phosphatase 2C ABI1 and the Ubiquitin-Proteasome Pathway. *Plant Cell Physiol* **56**(12), 2351-2367.

Miyazono K, Miyakawa T, Sawano Y, Kubota K, et al., 2009. Structural basis of abscisic acid signalling. *Nature* **462**(7273), 609-614.

Mizuno E, Kobayashi K, Yamamoto A, Kitamura N,Komada M, 2006. A deubiquitinating enzyme UBPY regulates the level of protein ubiquitination on endosomes. *Traffic* **7**(8), 1017-1031.

Moreno-Alvero M, Yunta C, Gonzalez-Guzman M, Lozano-Juste J, et al., 2017. Structure of Ligand-Bound Intermediates of Crop ABA Receptors Highlights PP2C as Necessary ABA Co-receptor. *Mol Plant* **10**(9), 1250-1253.

Morimoto K, Ohama N, Kidokoro S, Mizoi J, et al., 2017. BPM-CUL3 E3 ligase modulates thermotolerance by facilitating negative regulatory domain-mediated degradation of DREB2A in Arabidopsis. *Proc Natl Acad Sci U S A* **114**(40), E8528-e8536.

Mosquna A, Peterson F C, Park S Y, Lozano-Juste J, et al., 2011. Potent and selective activation of abscisic acid receptors in vivo by mutational stabilization of their agonist-bound conformation. *Proc Natl Acad Sci U S A* **108**(51), 20838-20843.

Mukhtar M S, Nishimura M T,Dangl J, 2009. NPR1 in Plant Defense: It's Not over 'til It's Turned over. *Cell* **137**(5), 804-806.

Mustilli A C, Merlot S, Vavasseur A, Fenzi F,Giraudat J, 2002. Arabidopsis OST1 protein kinase mediates the regulation of stomatal aperture by abscisic acid and acts upstream of reactive oxygen species production. *Plant Cell* **14**(12), 3089-3099.

Nakashima K, Fujita Y, Kanamori N, Katagiri T, et al., 2009. Three Arabidopsis SnRK2 protein kinases, SRK2D/SnRK2.2, SRK2E/SnRK2.6/OST1 and SRK2I/SnRK2.3, involved in ABA signaling are essential for the control of seed development and dormancy. *Plant Cell Physiol* **50**(7), 1345-1363.

Nambara E,Marion-Poll A, 2005. Abscisic acid biosynthesis and catabolism. *Annu Rev Plant Biol* **56**, 165-185.

Nee G, Kramer K, Nakabayashi K, Yuan B, et al., 2017. DELAY OF GERMINATION1 requires PP2C phosphatases of the ABA signalling pathway to control seed dormancy. *Nat Commun* **8**(1), 72.

Ng L M, Soon F F, Zhou X E, West G M, et al., 2011. Structural basis for basal activity and autoactivation of abscisic acid (ABA) signaling SnRK2 kinases. *Proc Natl Acad Sci U S A* **108**(52), 21259-21264.

Nishimura N, Sarkeshik A, Nito K, Park S Y, et al., 2010. PYR/PYL/RCAR family members are major in-vivo ABI1 protein phosphatase 2C-interacting proteins in Arabidopsis. *Plant J* **61**(2), 290-299.

Nishimura N, Tsuchiya W, Moresco J J, Hayashi Y, et al., 2018. Control of seed dormancy and germination by DOG1-AHG1 PP2C phosphatase complex via binding to heme. *Nat Commun* **9**(1), 2132.

Nishimura N, Yoshida T, Kitahata N, Asami T, et al., 2007. ABA-Hypersensitive Germination1 encodes a protein phosphatase 2C, an essential component of abscisic acid signaling in Arabidopsis seed. *Plant J* **50**(6), 935-949.

Ohta M, Guo Y, Halfter U,Zhu J K, 2003. A novel domain in the protein kinase SOS2 mediates interaction with the protein phosphatase 2C ABI2. *Proc Natl Acad Sci U S A* **100**(20), 11771-11776.

Ohtake F, Saeki Y, Sakamoto K, Ohtake K, et al., 2015. Ubiquitin acetylation inhibits polyubiquitin chain elongation. *EMBO Rep* **16**(2), 192-201.

Orman-Ligeza B, Morris E C, Parizot B, Lavigne T, et al., 2018. The Xerobranching Response Represses Lateral Root Formation When Roots Are Not in Contact with Water. *Curr Biol* **28**(19), 3165-3173.e3165.

Paez Valencia J, Goodman K,Otegui M S, 2016. Endocytosis and Endosomal Trafficking in Plants. *Annu Rev Plant Biol* **67**, 309-335.

Paiva S, Vieira N, Nondier I, Haguenauer-Tsapis R, et al., 2009. Glucose-induced ubiquitylation and endocytosis of the yeast Jen1 transporter: role of lysine 63-linked ubiquitin chains. *J Biol Chem* **284**(29), 19228-19236.

Park S Y, Fung P, Nishimura N, Jensen D R, et al., 2009. Abscisic acid inhibits type 2C protein phosphatases via the PYR/PYL family of START proteins. *Science* **324**(5930), 1068-1071.

Parry G, Calderon-Villalobos L I, Prigge M, Peret B, et al., 2009. Complex regulation of the TIR1/AFB family of auxin receptors. *Proc Natl Acad Sci U S A* **106**(52), 22540-22545.

Patra B, Pattanaik S,Yuan L, 2013. Ubiquitin protein ligase 3 mediates the proteasomal degradation of GLABROUS 3 and ENHANCER OF GLABROUS 3, regulators of trichome development and flavonoid biosynthesis in Arabidopsis. *Plant J* **74**(3), 435-447.

Pedmale U V,Liscum E, 2007. Regulation of phototropic signaling in Arabidopsis via phosphorylation state changes in the phototropin 1-interacting protein NPH3. *J Biol Chem* **282**(27), 19992-20001.

Peer W A, 2013. From perception to attenuation: auxin signalling and responses. *Curr Opin Plant Biol* **16**(5), 561-568.

Peirats-Llobet M, Han S K, Gonzalez-Guzman M, Jeong C W, et al., 2016. A Direct Link between Abscisic Acid Sensing and the Chromatin-Remodeling ATPase BRAHMA via Core ABA Signaling Pathway Components. *Mol Plant* **9**(1), 136-147.

Peters J M, 2006. The anaphase promoting complex/cyclosome: a machine designed to destroy. *Nat Rev Mol Cell Biol* **7**(9), 644-656.

Pickart C M,Fushman D, 2004. Polyubiquitin chains: polymeric protein signals. *Curr Opin Chem Biol* **8**(6), 610-616.

Pierre M, Traverso J A, Boisson B, Domenichini S, et al., 2007. N-myristoylation regulates the SnRK1 pathway in Arabidopsis. *Plant Cell* **19**(9), 2804-2821.

Pizzio G A, Rodriguez L, Antoni R, Gonzalez-Guzman M, et al., 2013. The PYL4 A194T mutant uncovers a key role of PYR1-LIKE4/PROTEIN PHOSPHATASE 2CA interaction for abscisic acid signaling and plant drought resistance. *Plant Physiol* **163**(1), 441-455.

Planes M D, Ninoles R, Rubio L, Bissoli G, et al., 2015. A mechanism of growth inhibition by abscisic acid in germinating seeds of Arabidopsis thaliana based on inhibition of plasma membrane H+-ATPase and decreased cytosolic pH, K+, and anions. *J Exp Bot* **66**(3), 813-825.

Potuschak T, Vansiri A, Binder B M, Lechner E, et al., 2006. The exoribonuclease XRN4 is a component of the ethylene response pathway in Arabidopsis. *Plant Cell* **18**(11), 3047-3057.

Prasad M E,Stone S L, 2010. Further analysis of XBAT32, an Arabidopsis RING E3 ligase, involved in ethylene biosynthesis. *Plant Signal Behav* **5**(11), 1425-1429.

Qiao H, Chang K N, Yazaki J,Ecker J R, 2009. Interplay between ethylene, ETP1/ETP2 F-box proteins, and degradation of EIN2 triggers ethylene responses in Arabidopsis. *Genes Dev* **23**(4), 512-521.

Radauer C, Lackner P,Breiteneder H, 2008. The Bet v 1 fold: an ancient, versatile scaffold for binding of large, hydrophobic ligands. *BMC Evol Biol* **8**, 286.

Ren X, Chen Z, Liu Y, Zhang H, et al., 2010. ABO3, a WRKY transcription factor, mediates plant responses to abscisic acid and drought tolerance in Arabidopsis. *Plant J* **63**(3), 417-429.

Reyes F C, Buono R A, Roschzttardtz H, Di Rubbo S, et al., 2014. A novel endosomal sorting complex required for transport (ESCRT) component in Arabidopsis thaliana controls cell expansion and development. *J Biol Chem* **289**(8), 4980-4988.

Richardson L G, Howard A S, Khuu N, Gidda S K, et al., 2011. Protein-Protein Interaction Network and Subcellular Localization of the Arabidopsis Thaliana ESCRT Machinery. *Front Plant Sci* **2**, 20.

Riley B E, Lougheed J C, Callaway K, Velasquez M, et al., 2013. Structure and function of Parkin E3 ubiquitin ligase reveals aspects of RING and HECT ligases. *Nat Commun* **4**, 1982.

Rodrigues A, Adamo M, Crozet P, Margalha L, et al., 2013. ABI1 and PP2CA phosphatases are negative regulators of Snf1-related protein kinase1 signaling in Arabidopsis. *Plant Cell* **25**(10), 3871-3884.

Rodriguez L, Gonzalez-Guzman M, Diaz M, Rodrigues A, et al., 2014. C2-domain abscisic acid-related proteins mediate the interaction of PYR/PYL/RCAR abscisic acid receptors with the plasma membrane and regulate abscisic acid sensitivity in Arabidopsis. *Plant Cell* **26**(12), 4802-4820.

Rodriguez P L, Benning G,Grill E, 1998. ABI2, a second protein phosphatase 2C involved in abscisic acid signal transduction in Arabidopsis. *FEBS Lett* **421**(3), 185-190.

Romero-Barrios N,Vert G, 2018. Proteasome-independent functions of lysine-63 polyubiquitination in plants. *New Phytol* **217**(3), 995-1011.

Ron M, Alandete Saez M, Eshed Williams L, Fletcher J C,McCormick S, 2010. Proper regulation of a sperm-specific cis-nat-siRNA is essential for double fertilization in Arabidopsis. *Genes Dev* **24**(10), 1010-1021.

Row P E, Prior I A, McCullough J, Clague M J,Urbe S, 2006. The ubiquitin isopeptidase UBPY regulates endosomal ubiquitin dynamics and is essential for receptor down-regulation. *J Biol Chem* **281**(18), 12618-12624.

Rubio S, Rodrigues A, Saez A, Dizon M B, et al., 2009. Triple loss of function of protein phosphatases type 2C leads to partial constitutive response to endogenous abscisic acid. *Plant Physiol* **150**(3), 1345-1355.

Ryu M Y, Cho S K, Hong Y, Kim J, et al., 2019. Classification of barley U-box E3 ligases and their expression patterns in response to drought and pathogen stresses. *BMC Genomics* **20**(1), 326.

Saez A, Apostolova N, Gonzalez-Guzman M, Gonzalez-Garcia M P, et al., 2004. Gain-of-function and loss-of-function phenotypes of the protein phosphatase 2C HAB1 reveal its role as a negative regulator of abscisic acid signalling. *Plant J* **37**(3), 354-369.

Saez A, Robert N, Maktabi M H, Schroeder J I, et al., 2006. Enhancement of abscisic acid sensitivity and reduction of water consumption in Arabidopsis by combined inactivation of the protein phosphatases type 2C ABI1 and HAB1. *Plant Physiol* **141**(4), 1389-1399.

Saez A, Rodrigues A, Santiago J, Rubio S,Rodriguez P L, 2008. HAB1-SWI3B interaction reveals a link between abscisic acid signaling and putative SWI/SNF chromatin-remodeling complexes in Arabidopsis. *Plant Cell* **20**(11), 2972-2988.

Salazar-Cerezo S, Martinez-Montiel N, Garcia-Sanchez J, Perez Y T R,Martinez-Contreras R D, 2018. Gibberellin biosynthesis and metabolism: A convergent route for plants, fungi and bacteria. *Microbiol Res* **208**, 85-98.

Santiago J, Dupeux F, Betz K, Antoni R, et al., 2012. Structural insights into PYR/PYL/RCAR ABA receptors and PP2Cs. *Plant Sci* **182**, 3-11.

Santiago J, Rodrigues A, Saez A, Rubio S, et al., 2009. Modulation of drought resistance by the abscisic acid receptor PYL5 through inhibition of clade A PP2Cs. *Plant J* **60**(4), 575-588.

Sato A, Sato Y, Fukao Y, Fujiwara M, et al., 2009a. Threonine at position 306 of the KAT1 potassium channel is essential for channel activity and is a target site for ABA-activated SnRK2/OST1/SnRK2.6 protein kinase. *Biochem J* **424**(3), 439-448.

Sato Y, Yoshikawa A, Mimura H, Yamashita M, et al., 2009b. Structural basis for specific recognition of Lys 63-linked polyubiquitin chains by tandem UIMs of RAP80. *Embo j* **28**(16), 2461-2468.

Schweighofer A, Hirt H,Meskiene I, 2004. Plant PP2C phosphatases: emerging functions in stress signaling. *Trends Plant Sci* **9**(5), 236-243.

Scott D C, Rhee D Y, Duda D M, Kelsall I R, et al., 2016. Two Distinct Types of E3 Ligases Work in Unison to Regulate Substrate Ubiquitylation. *Cell* **166**(5), 1198-1214.e1124.

Seo D H, Ryu M Y, Jammes F, Hwang J H, et al., 2012. Roles of four Arabidopsis U-box E3 ubiquitin ligases in negative regulation of abscisic acid-mediated drought stress responses. *Plant Physiol* **160**(1), 556-568.

Seo H S, Yang J Y, Ishikawa M, Bolle C, et al., 2003. LAF1 ubiquitination by COP1 controls photomorphogenesis and is stimulated by SPA1. *Nature* **423**(6943), 995-999.

Shang Y, Yan L, Liu Z Q, Cao Z, et al., 2010. The Mg-chelatase H subunit of Arabidopsis antagonizes a group of WRKY transcription repressors to relieve ABA-responsive genes of inhibition. *Plant Cell* **22**(6), 1909-1935.

Sheard L B, Tan X, Mao H, Withers J, et al., 2010. Jasmonate perception by inositol-phosphate-potentiated COI1-JAZ co-receptor. *Nature* **468**(7322), 400-405.

Sheen J, 1996. Ca2+-dependent protein kinases and stress signal transduction in plants. *Science* **274**(5294), 1900-1902.

Shen J, Gao C, Zhao Q, Lin Y, et al., 2016. AtBRO1 Functions in ESCRT-I Complex to Regulate Multivesicular Body Protein Sorting. *Mol Plant* **9**(5), 760-763.

Shih S C, Katzmann D J, Schnell J D, Sutanto M, et al., 2002. Epsins and Vps27p/Hrs contain ubiquitin-binding domains that function in receptor endocytosis. *Nat Cell Biol* **4**(5), 389-393.

Shih S C, Sloper-Mould K E,Hicke L, 2000. Monoubiquitin carries a novel internalization signal that is appended to activated receptors. *Embo j* **19**(2), 187-198.

Shimada A, Ueguchi-Tanaka M, Nakatsu T, Nakajima M, et al., 2008. Structural basis for gibberellin recognition by its receptor GID1. *Nature* **456**(7221), 520-523.

Signora L, De Smet I, Foyer C H,Zhang H, 2001. ABA plays a central role in mediating the regulatory effects of nitrate on root branching in Arabidopsis. *Plant J* **28**(6), 655-662.

Sims J J,Cohen R E, 2009. Linkage-specific avidity defines the lysine 63-linked polyubiquitin-binding preference of rap80. *Mol Cell* **33**(6), 775-783.

Sims J J, Haririnia A, Dickinson B C, Fushman D,Cohen R E, 2009. Avid interactions underlie the Lys63-linked polyubiquitin binding specificities observed for UBA domains. *Nat Struct Mol Biol* **16**(8), 883-889.

Sirichandra C, Davanture M, Turk B E, Zivy M, et al., 2010. The Arabidopsis ABA-activated kinase OST1 phosphorylates the bZIP transcription factor ABF3 and creates a 14-3-3 binding site involved in its turnover. *PLoS One* **5**(11), e13935.

Skowyra D, Craig K L, Tyers M, Elledge S J,Harper J W, 1997. F-box proteins are receptors that recruit phosphorylated substrates to the SCF ubiquitin-ligase complex. *Cell* **91**(2), 209-219.

Sloper-Mould K E, Jemc J C, Pickart C M,Hicke L, 2001. Distinct functional surface regions on ubiquitin. *J Biol Chem* **276**(32), 30483-30489.

Soderman E M, Brocard I M, Lynch T J,Finkelstein R R, 2000. Regulation and function of the Arabidopsis ABA-insensitive4 gene in seed and abscisic acid response signaling networks. *Plant Physiol* **124**(4), 1752-1765.

Soon F F, Ng L M, Zhou X E, West G M, et al., 2012. Molecular mimicry regulates ABA signaling by SnRK2 kinases and PP2C phosphatases. *Science* **335**(6064), 85-88.

Soucy T A, Smith P G, Milhollen M A, Berger A J, et al., 2009. An inhibitor of NEDD8-activating enzyme as a new approach to treat cancer. *Nature* **458**(7239), 732-736.

Spoel S H, Mou Z, Tada Y, Spivey N W, et al., 2009. Proteasome-mediated turnover of the transcription coactivator NPR1 plays dual roles in regulating plant immunity. *Cell* **137**(5), 860-872.

Stone S L, Hauksdottir H, Troy A, Herschleb J, et al., 2005. Functional analysis of the RING-type ubiquitin ligase family of Arabidopsis. *Plant Physiol* **137**(1), 13-30.

Swaney D L, Rodriguez-Mias R A,Villen J, 2015. Phosphorylation of ubiquitin at Ser65 affects its polymerization, targets, and proteome-wide turnover. *EMBO Rep* **16**(9), 1131-1144.

Szostkiewicz I, Richter K, Kepka M, Demmel S, et al., 2010. Closely related receptor complexes differ in their ABA selectivity and sensitivity. *Plant J* **61**(1), 25-35.

Tan X, Calderon-Villalobos L I, Sharon M, Zheng C, et al., 2007. Mechanism of auxin perception by the TIR1 ubiquitin ligase. *Nature* **446**(7136), 640-645.

Tanno H,Komada M, 2013. The ubiquitin code and its decoding machinery in the endocytic pathway. *J Biochem* **153**(6), 497-504.

Tasaki T, Sriram S M, Park K S,Kwon Y T, 2012. The N-end rule pathway. *Annu Rev Biochem* **81**, 261-289.

Tenno T, Fujiwara K, Tochio H, Iwai K, et al., 2004. Structural basis for distinct roles of Lys63- and Lys48-linked polyubiquitin chains. *Genes Cells* **9**(10), 865-875.

Thines B, Parlan E V,Fulton E C, 2019. Circadian Network Interactions with Jasmonate Signaling and Defense. *Plants (Basel)* **8**(8).

Tian M,Xie Q, 2013. Non-26S proteasome proteolytic role of ubiquitin in plant endocytosis and endosomal trafficking(F). *J Integr Plant Biol* **55**(1), 54-63.

Tischer S V, Wunschel C, Papacek M, Kleigrewe K, et al., 2017. Combinatorial interaction network of abscisic acid receptors and coreceptors from Arabidopsis thaliana. *Proc Natl Acad Sci U S A* **114**(38), 10280-10285.

Turnbull D,Hemsley P A, 2017. Fats and function: protein lipid modifications in plant cell signalling. *Curr Opin Plant Biol* **40**, 63-70.

Umezawa T, Sugiyama N, Mizoguchi M, Hayashi S, et al., 2009. Type 2C protein phosphatases directly regulate abscisic acid-activated protein kinases in Arabidopsis. *Proc Natl Acad Sci U S A* **106**(41), 17588-17593.

Uno Y, Furihata T, Abe H, Yoshida R, et al., 2000. Arabidopsis basic leucine zipper transcription factors involved in an abscisic acid-dependent signal transduction pathway under drought and high-salinity conditions. *Proc Natl Acad Sci U S A* **97**(21), 11632-11637.

Varshavsky A, 2011. The N-end rule pathway and regulation by proteolysis. *Protein Sci* **20**(8), 1298-1345.

Varshavsky A, 2019. N-degron and C-degron pathways of protein degradation. *Proc Natl Acad Sci U S A* **116**(2), 358-366.

Vlad F, Rubio S, Rodrigues A, Sirichandra C, et al., 2009. Protein phosphatases 2C regulate the activation of the Snf1-related kinase OST1 by abscisic acid in Arabidopsis. *Plant Cell* **21**(10), 3170-3184.

Waadt R, Hitomi K, Nishimura N, Hitomi C, et al., 2014. FRET-based reporters for the direct visualization of abscisic acid concentration changes and distribution in Arabidopsis. *Elife* **3**, e01739.

Walter M, Chaban C, Schutze K, Batistic O, et al., 2004. Visualization of protein interactions in living plant cells using bimolecular fluorescence complementation. *Plant J* **40**(3), 428-438.

Wang K, He J, Zhao Y, Wu T, et al., 2018a. EAR1 Negatively Regulates ABA Signaling by Enhancing 2C Protein Phosphatase Activity. *Plant Cell* **30**(4), 815-834.

Wang K L, Yoshida H, Lurin C,Ecker J R, 2004. Regulation of ethylene gas biosynthesis by the Arabidopsis ETO1 protein. *Nature* **428**(6986), 945-950.

Wang P, Zhao Y, Li Z, Hsu C C, et al., 2018b. Reciprocal Regulation of the TOR Kinase and ABA Receptor Balances Plant Growth and Stress Response. *Mol Cell* **69**(1), 100-112.e106.

Wang P, Zheng Y, Guo Y, Chen X, et al., 2019a. Identification, expression, and putative target gene analysis of nuclear factor-Y (NF-Y) transcription factors in tea plant (Camellia sinensis). *Planta*.

Wang X, Guo C, Peng J, Li C, et al., 2019b. ABRE-BINDING FACTORS play a role in the feedback regulation of ABA signaling by mediating rapid ABA induction of ABA co-receptor genes. *New Phytol* **221**(1), 341-355.

Webb A A, Larman M G, Montgomery L T, Taylor J E,Hetherington A M, 2001. The role of calcium in ABA-induced gene expression and stomatal movements. *Plant J* **26**(3), 351-362.

Weber H, Bernhardt A, Dieterle M, Hano P, et al., 2005. Arabidopsis AtCUL3a and AtCUL3b form complexes with members of the BTB/POZ-MATH protein family. *Plant Physiol* **137**(1), 83-93.

Weng J K, Ye M, Li B,Noel J P, 2016. Co-evolution of Hormone Metabolism and Signaling Networks Expands Plant Adaptive Plasticity. *Cell* **166**(4), 881-893.

Wenzel D M,Klevit R E, 2012. Following Ariadne's thread: a new perspective on RBR ubiquitin ligases. *BMC Biol* **10**, 24.

Wenzel D M, Lissounov A, Brzovic P S,Klevit R E, 2011. UBCH7 reactivity profile reveals parkin and HHARI to be RING/HECT hybrids. *Nature* **474**(7349), 105-108.

Winklhofer K F, 2014. Parkin and mitochondrial quality control: toward assembling the puzzle. *Trends Cell Biol* **24**(6), 332-341.

Winter D, Vinegar B, Nahal H, Ammar R, et al., 2007. An "Electronic Fluorescent Pictograph" browser for exploring and analyzing large-scale biological data sets. *PLoS One* **2**(8), e718.

Winter V,Hauser M T, 2006. Exploring the ESCRTing machinery in eukaryotes. *Trends Plant Sci* **11**(3), 115-123.

Woo O G, Kim S H, Cho S K, Kim S H, et al., 2018. BPH1, a novel substrate receptor of CRL3, plays a repressive role in ABA signal transduction. *Plant Mol Biol* **96**(6), 593-606.

Wu P Y, Hanlon M, Eddins M, Tsui C, et al., 2003. A conserved catalytic residue in the ubiquitin-conjugating enzyme family. *Embo j* **22**(19), 5241-5250.

Wu Q, Zhang X, Peirats-Llobet M, Belda-Palazon B, et al., 2016. Ubiquitin Ligases RGLG1 and RGLG5 Regulate Abscisic Acid Signaling by Controlling the Turnover of Phosphatase PP2CA. *Plant Cell* **28**(9), 2178-2196.

Xu R Y, Xu J, Wang L, Niu B, et al., 2019. The Arabidopsis anaphase-promoting complex/cyclosome subunit 8 is required for male meiosis. *New Phytol.*

Yanagawa Y, Sullivan J A, Komatsu S, Gusmaroli G, et al., 2004. Arabidopsis COP10 forms a complex with DDB1 and DET1 in vivo and enhances the activity of ubiquitin conjugating enzymes. *Genes Dev* **18**(17), 2172-2181.

Yang Y, Sulpice R, Himmelbach A, Meinhard M, et al., 2006. Fibrillin expression is regulated by abscisic acid response regulators and is involved in abscisic acid-mediated photoprotection. *Proc Natl Acad Sci U S A* **103**(15), 6061-6066.

Yang Z, Liu J, Poree F, Schaeufele R, et al., 2019. Abscisic Acid Receptors and Coreceptors Modulate Plant Water Use Efficiency and Water Productivity. *Plant Physiol* **180**(2), 1066-1080.

Yee D,Goring D R, 2009. The diversity of plant U-box E3 ubiquitin ligases: from upstream activators to downstream target substrates. *J Exp Bot* **60**(4), 1109-1121.

Yin P, Fan H, Hao Q, Yuan X, et al., 2009. Structural insights into the mechanism of abscisic acid signaling by PYL proteins. *Nat Struct Mol Biol* **16**(12), 1230-1236.

Yin X J, Volk S, Ljung K, Mehlmer N, et al., 2007. Ubiquitin lysine 63 chain forming ligases regulate apical dominance in Arabidopsis. *Plant Cell* **19**(6), 1898-1911.

Yoo S D, Cho Y H,Sheen J, 2007. Arabidopsis mesophyll protoplasts: a versatile cell system for transient gene expression analysis. *Nat Protoc* **2**(7), 1565-1572.

Yoshida R, Hobo T, Ichimura K, Mizoguchi T, et al., 2002. ABA-activated SnRK2 protein kinase is required for dehydration stress signaling in Arabidopsis. *Plant Cell Physiol* **43**(12), 1473-1483.

Yoshida T, Christmann A, Yamaguchi-Shinozaki K, Grill E,Fernie A R, 2019. Revisiting the Basal Role of ABA - Roles Outside of Stress. *Trends Plant Sci* **24**(7), 625-635.

Yoshida T, Fujita Y, Maruyama K, Mogami J, et al., 2015. Four Arabidopsis AREB/ABF transcription factors function predominantly in gene expression downstream of SnRK2 kinases in abscisic acid signalling in response to osmotic stress. *Plant Cell Environ* **38**(1), 35-49.

Yoshida T, Fujita Y, Sayama H, Kidokoro S, et al., 2010. AREB1, AREB2, and ABF3 are master transcription factors that cooperatively regulate ABRE-dependent ABA signaling involved in drought stress tolerance and require ABA for full activation. *Plant J* **61**(4), 672-685.

Yoshida T, Nishimura N, Kitahata N, Kuromori T, et al., 2006. ABA-hypersensitive germination3 encodes a protein phosphatase 2C (AtPP2CA) that strongly regulates abscisic acid signaling during germination among Arabidopsis protein phosphatase 2Cs. *Plant Physiol* **140**(1), 115-126.

Yu F, Lou L, Tian M, Li Q, et al., 2016. ESCRT-I Component VPS23A Affects ABA Signaling by Recognizing ABA Receptors for Endosomal Degradation. *Mol Plant* **9**(12), 1570-1582.

Yu F, Qian L, Nibau C, Duan Q, et al., 2012. FERONIA receptor kinase pathway suppresses abscisic acid signaling in Arabidopsis by activating ABI2 phosphatase. *Proc Natl Acad Sci U S A* **109**(36), 14693-14698.

Yu F,Xie Q, 2017. Non-26S Proteasome Endomembrane Trafficking Pathways in ABA Signaling. *Trends Plant Sci* **22**(11), 976-985.

Yu Z, Zhang D, Xu Y, Jin S, et al., 2019. CEPR2 phosphorylates and accelerates the degradation of PYR/PYLs in Arabidopsis. *J Exp Bot*.

Zentgraf U, Laun T,Miao Y, 2010. The complex regulation of WRKY53 during leaf senescence of Arabidopsis thaliana. *Eur J Cell Biol* **89**(2-3), 133-137.

Zhang H, Gannon L, Jones P D, Rundle C A, et al., 2018a. Genetic interactions between ABA signalling and the Arg/N-end rule pathway during Arabidopsis seedling establishment. *Sci Rep* **8**(1), 15192.

Zhang J, Ren Z, Zhou Y, Ma Z, et al., 2019. NPR1 and Redox Rhythmx: Connections, between Circadian Clock and Plant Immunity. *Int J Mol Sci* **20**(5).

Zhang L, Li X, Li D, Sun Y, et al., 2018b. CARK1 mediates ABA signaling by phosphorylation of ABA receptors. *Cell Discov* **4**, 30.

Zhang X, Wang N, Chen P, Gao M, et al., 2014. Overexpression of a soybean ariadne-like ubiquitin ligase gene GmARI1 enhances aluminum tolerance in Arabidopsis. *PLoS One* **9**(11), e111120.

Zhang X, Wu Q, Ren J, Qian W, et al., 2012. Two novel RING-type ubiquitin ligases, RGLG3 and RGLG4, are essential for jasmonate-mediated responses in Arabidopsis. *Plant Physiol* **160**(2), 808-822.

Zhao J, Zhao L, Zhang M, Zafar S A, et al., 2017. Arabidopsis E3 Ubiquitin Ligases PUB22 and PUB23 Negatively Regulate Drought Tolerance by Targeting ABA Receptor PYL9 for Degradation. *Int J Mol Sci* **18**(9).

Zhao Q, Tian M, Li Q, Cui F, et al., 2013. A plant-specific in vitro ubiquitination analysis system. *Plant J* **74**(3), 524-533.

Zhao Y, Chan Z, Gao J, Xing L, et al., 2016. ABA receptor PYL9 promotes drought resistance and leaf senescence. *Proc Natl Acad Sci U S A* **113**(7), 1949-1954.

Zhao Y, Xing L, Wang X, Hou Y J, et al., 2014. The ABA receptor PYL8 promotes lateral root growth by enhancing MYB77-dependent transcription of auxin-responsive genes. *Sci Signal* **7**(328), ra53.

Zhao Y, Zhang Z, Gao J, Wang P, et al., 2018. Arabidopsis Duodecuple Mutant of PYL ABA Receptors Reveals PYL Repression of ABA-Independent SnRK2 Activity. *Cell Rep* **23**(11), 3340-3351.e3345.

Zheng N, Schulman B A, Song L, Miller J J, et al., 2002. Structure of the Cul1-Rbx1-Skp1-F boxSkp2 SCF ubiquitin ligase complex. *Nature* **416**(6882), 703-709.

Zhou H, Zhao J, Cai J,Patil S B, 2017. UBIQUITIN-SPECIFIC PROTEASES function in plant development and stress responses. *Plant Mol Biol* **94**(6), 565-576.

Zhu M, Zhang T, Ji W, Silva-Sanchez C, et al., 2017. Redox regulation of a guard cell SNF1-related protein kinase in Brassica napus, an oilseed crop. *Biochem J* **474**(15), 2585-2599.

Zhuang M, Calabrese M F, Liu J, Waddell M B, et al., 2009. Structures of SPOP-substrate complexes: insights into molecular architectures of BTB-Cul3 ubiquitin ligases. *Mol Cell* **36**(1), 39-50.

ABBREVIATIONS

ABA	:	abscisic acid
ABI1	:	ABA INSENSITIVE 1
ABF	:	ABRE binding factor
ABRE	:	ABA responsive element
ACC	:	1-aminocyclopropane-1-carboxylic acid
ACS	:	ACC SYNTHASE
ACT	:	actin
AD	:	Gal4 activation domain
AEL	:	ARABIDOPSIS EL1-LIKE
AFB	:	AUXIN-BINDING FBX
AHG1	:	ABA HYPERSENSITIVE GERMINATION 1
ALIX	:	ALG-2 INTERACTING PROTEIN-X
APC/C	:	ANAPHASE PROMOTING COMPLEX/CYCLOSOME
AP2	:	APETALA2
APS	:	ammonium persulfate
AREB	:	ABRE-binding proteins
ARF	:	auxin response factor
ARI	:	Ariadne
Asp	:	aspartic acid
AtDPBF	:	Dc3 PROMOTER BINDING FACTOR
ATL	:	Arabidopsis genes toxic to yeast
AtPP2-B11	:	PHLOEM PROTEIN 2-B11
Aux/IAA	:	auxin/indole-3-acetic acid
BD	:	Gal4 DNA binding domain

BiFC	:	bimolecular fluorescence complementation
BIN2	:	BRASSINOSTEROID INSENSITIVE 2
BL	:	brassinolide
BPH1	:	BTB/POZ PROTEIN HYPERSENSITIVE TO ABA 1
BTB/POZ	:	Broad-complex, Tramtrack, and Bric-à-brac / POx virus and Zinc finger
CAND	:	CULLIN-ASSOCIATED NEDD8-DISSOCIATED
CAR	:	C2-DOMAIN ABA-RELATED
CBL	:	calcineurin B-like
CDF	:	CYCLIN DOF FACTOR
CEPR2	:	C-terminally encoded peptide receptor 2
CHIP	:	CARBOXYL TERMINUS OF HSC70-INTERACTING PROTEIN
CHX:	:	cycloheximide
CIPK	:	CBL-interacting protein kinases
coIP	:	coimmunoprecipitation
COP	:	CONSTITUTIVELY PHOTOMORPHOGENIC
CRL	:	cullin-RING based E3 ligase
CSN	:	COP9/signalosome
CTAB	:	cetyl trimethyl ammonium bromide
CTR	:	CONSTITUTIVE TRIPLE RESPONSE
CUL	:	cullin
d	:	days
DAPI	:	4',6-diamidino-2-phenylindole

DDA1	:	DDB1-ASSOCIATED1
DDB	:	DNA damage-binding protein
DET	:	DE-ETIOLATED
DOG1	:	DELAY OF GERMINATION 1
DRE	:	drought response elements
DTT	:	dithiothreitol
DUB	:	deubiquitylating enzyme
DWD	:	DDB1-binding/WD-40 domain containing
EAR1	:	ENHANCER OF ABA CO-RECEPTOR 1
EBF	:	EIN3-binding FBX protein
EGL3	:	ENHANCER OF GL3
EGR2	:	CLADE-E GROWTH-REGULATING 2
EIL	:	EIN3-LIKE
EIN	:	ETHYLENE-INSENSITIVE
EOL	:	ETO1-LIKE
ESCRT	:	endosomal sorting complex required for transport
ETO	:	ETHYLENE OVERPRODUCER
ETP	:	EIN2-targeting protein
ETR	:	ETHYLENE RESPONSE
FBX	:	F-box
FER	:	Feronia
Fib	:	Fibrillarin
FKF1	:	FLAVIN BINDING KELCH REPEAT F-BOX1
FLS2	:	FLAGELLIN SENSING2

FREE1	:	FYVE DOMAIN PROTEIN REQUIRED FOR ENDOSOMAL SORTING 1
FYVE1	:	FYVE DOMAIN-CONTAINING PROTEIN 1
GA	:	gibberellin
GEF1	:	guanine exchange factor 1
GI	:	GIGANTEA
GID1	:	GIBBERELLIN INSENSITIVE DWARF1
GL3	:	GLABROUS 3
Gly	:	glycine
GSK3	:	ARABIDOPSIS GLYCOGEN SYNTHASE KINASE 3
G_{ST}	:	stomatal conductance
h	:	hours
HAB1	:	HYPERSENSITIVE TO ABA 1
HAI1	:	HIGHLY ABA INDUCIBLE 1
HHARI	:	Human Homologue of Ariadne
HRP	:	horseradish peroxidase
IAA	:	indole-3-acetic acid
IBR	:	In-between RING
ILV	:	intraluminal vesicles
IP	:	immunoprecipitation
JA	:	jasmonate
JA-Ile	:	jasmonoyl isoleucine
KAT1	:	POTASSIUM CHANNEL IN ARABIDOPSIS THALIANA 1

LRB1	:	LIGHT-RESPONSE BTB 1
LC-MS/MS	:	liquid chromatography-tandem mass spectrometry
LKP2	:	LOV KELCH PROTEIN 2
LOV	:	Light, Oxygen, or Voltage
LRR-RLK	:	Leucine-rich repeat receptor-like kinase
LUC	:	luciferase
MATH	:	meprin-and-TRAF-homology
min	:	minutes
MJ	:	methyl jasmonate
MS	:	Murashige–Skoog
MUG	:	4-methyl umbelliferyl glucuronide
NEM	:	N-ethylmaleimide
NLS	:	nuclear localization site
NMT1	:	N-myristoyltransferase 1
NO	:	nitric oxide
NPH3	:	nonphototropic-hypocotyl 3
NPR1	:	Non-expresser of PR genes 1
OE	:	overexpressing lines
PA	:	phaseic acid
PD	:	pull-down
PP1	:	type 1 phosphatases
PP2	:	type 2 phosphatases
PP2C	:	protein phosphatase type 2C
PP2CA	:	PROTEIN PHOSPHATASE 2CA

PRR	:	pseudoresponse regulator
PUB	:	plant U-box
PVDF	:	polyvinylidene difluoride
PYL	:	PYR1-LIKE
PYR1	:	PYRABACTIN RESISTANCE1
QUAC1	:	R-type ion channel 1
RALF	:	rapid alkalinisation factor
RBR	:	RING between RING
RBX1	:	RING BOX 1
RCAR	:	REGULATORY COMPONENTS OF ABA RECEPTORS
RDK1	:	RECEPTOR DEAD KINASE 1
RFA	:	RING finger ABA-related
RGLG1	:	RING DOMAIN LIGASE 1
RING	:	REALLY INTERESTING NEW GEN
RLCK	:	RECEPTOR-LIKE CYTOPLASMIC KINASE
ROS	:	reactive oxygen species
ROX	:	carboxy-X-rhodamine
R_P	:	Pearson's correlation coefficient
RSL1	:	RBR-type RING FINGER OF SEED LONGEVITY
RUB	:	RELATED TO UB
SCD	:	synthetic complete defined
SCF	:	SKP1-Cullin-1-F-box protein
SD	:	synthetic defined

SDS-PAGE	:	sodium dodecyl sulfate polyacrylamide gel electrophoresis
Ser	:	serine
SKP	:	S-phase kinase-assiciated protein
SLAC1	:	SLOW ANION CHANNEL 1
SLY	:	SLEEPY
SNE	:	SNEEZY
SnRK2	:	ABA-activated SNF1-related protein kinases
S_P	:	Spearman's correlation coefficient
SPA	:	SUPPRESSOR OF PHYA
Split-LUC	:	split-luciferase
SP1	:	SUPPRESSOR OF PLASTID PROTEIN IMPORT LOCUS
TE	:	Tiller Enhancer
TEMED	:	N,N,N',N'-tetramethyl ethylenediamine
TF	:	transcription factor
Thr	:	threonine
TOC	:	Translocation at the outer envelope of chloroplasts
TOC1	:	TIMING OF CAB EXPRESSION 1
TIR	:	TRANSPORT INHIBITOR RESPONSE
TOR	:	TARGET OF RAMPAMYCIN
Trp	:	tryptophan
Tyr	:	tyrosine
UBA1	:	UBIQUITIN ACTIVATING 1

UBC	:	ubiquitin conjugating
UBP	:	UBIQUITIN-SPECIFIC PROTEASES
UPL	:	UBIQUITIN PROTEIN LIGASES
VPS23A	:	VACUOLAR PROTEIN SORTING 23A
w	:	weeks
WUE	:	water use efficiency
Y2H	:	yeast two hybrid
YNB	:	yeast nitrogen base
ZTL	:	ZEITLUPE

APPENDIX

Treatment	Protein name	Peptides identified by MS	Peptide score/ expect p
Mock	PP2CA	VLGVLAMSR	45.56/0.00017
		VLGVLAMSR	44.05/0.00024
		VLGVLAMSR	40.57/0.00053
		VIYWDGAR	34.86/0.0027
		RLDLLPSIK	27.61/0.0049
		QSSDNVSVVVVDLR	29.04/0.012
		QSSDNVSVVVVDLR	26.91/0.02
		QSSDNVSVVVVDLR	52.89/5.80E-05
MG-132 + ABA 12h	PP2CA	VLGVLAMSR	45,27/0,00018
		VLGVLAMSR	36,74/0,0021
		VLGVLAMSR	23,42/0,045
		VIYWDGAR	24,37/0,031
		QSSDNVSVVVVDLR	49,59/0,00011
		QSSDNVSVVVVDLR	65,95/2,40E-06
		QSSDNVSVVVVDLR	47,47/0,0002
		IVADSAVAPPLENCR	59,3/1,20E-05
		QSSDNVSVVVVDLRK	49,19/9,60E-05
		QSSDNVSVVVVDLRK	42,48/0,00055
		QSSDNVSVVVVDLRK	45,82/0,00026
	BPM3	AQFFGPIGNNNVDR	41,51/0,00053
MG-132 24h	PP2CA	VLGVLAMSR	64,78/2,00E-06
		VIYWDGAR	24,34/0,03
		VIYWDGAR	25,42/0,024
		RLDLLPSIK	45,74/7,50E-05
		RLDLLPSIK	34,07/0,0011
		QSSDNVSVVVVDLR	68,9/1,20E-06
		IVADSAVAPPLENCR	46,78/0,00022
		IVADSAVAPPLENCR	22,17/0,06
		IVADSAVAPPLENCR	36,54/0,0022
		DMEDAVSIHPSFLQR	32,15/0,0032
		DMEDAVSIHPSFLQR	65,4/1,70E-06
		GAGAGDDSDAAHNACSDAALLLTK	57,49/8,20E-06
		GAGAGDDSDAAHNACSDAALLLTK	40,34/0,00045
	BPM5	SQSVWAQLSDGGGDTTSR	52,94/1,50E-05

Treatment	Protein name	Peptides identified by MS	Peptide score/ expect p
MG-132 24h +ABA 6h (I)	PP2CA	VLGVLAMSR	57,14/1,20E-05
		VIYWDGAR	23,11/0,042
		VIYWDGAR	24,92/0,027
		RLDLLPSIK	24,79/0,0093
		RLDLLPSIK	21,11/0,022
		RLDLLPSIK	47,57/4,90E-05
		QSSDNVSVVVVDLR	67,23/1,80E-06
		QSSDNVSVVVVDLR	42,34/0,00056
		IVADSAVAPPLENCR	58,35/1,30E-05
		IVADSAVAPPLENCR	40,92/0,0008
		QSSDNVSVVVVDLRK	42,78/0,00051
		QSSDNVSVVVVDLRK	41,29/0,00062
		DMEDAVSIHPSFLQR	24,03/0,018
		DMEDAVSIHPSFLQR	23,06/0,023
		DMEDAVSIHPSFLQR	24,13/0,019
		DMEDAVSIHPSFLQR	45,6/0,00015
		RRDMEDAVSIHPSFLQR	35,7/0,0022
		GAGAGDDSDAAHNACSDAALLLTK	45,08/0,00013
		GAGAGDDSDAAHNACSDAALLLTK	55,88/1,20E-05
	BPM5	SQSVWAQLSDGGGDTTSR	67,09/6,00E-07
MG-132 24h +ABA 6h (II)	PP2CA	VLGVLAMSR	53.72/2.60E-05
		VLGVLAMSR	52.81/3.20E-05
		VLGVLAMSR	40.73/0.00081
		VIYWDGAR	31.8/0.0055
		RLDLLPSIK	30.99/0.0022
		RLDLLPSIK	33.59/0.0012
		ECNLVVNGATR	48.42/8.80E-05
		IVADSAVAPPLENCR	56.49/2.30E-05
		QSSDNVSVVVVDLRK	23.53/0.035
		DMEDAVSIHPSFLQR	44.75/0.00015
		DMEDAVSIHPSFLQR	27.49/0.0089
		MDKEVSQRECNLVVNGATR	26.01/0.017
		GAGAGDDSDAAHNACSDAALLLTK	37.14/0.00079
		GAGAGDDSDAAHNACSDAALLLTK	69.76/4.90E-07
	BPM3	SASSVLGCDTTNVR	38.88/0.0008
		AQFFGPIGNNNVDR	29.74/0.0082
	BPM4	ALESGPYTLK	20.2/0.092
		ALESGPYTLK	34.27/0.0038
		TVAGCEEELSGGGGK	62.89/1.60E-06
	BPM5	SLDGGPYTLK	44.95/0.00032
		SQSVWAQLSDGGGDTTSR	90.3/3,00E-09
		SQSVWAQLSDGGGDTTSR	71.99/2,10E-07

Treatment	Protein name	Peptides identified by MS	Peptide score/ expect p
MG-132 24h +ABA 6h (II)	RGLG1	IPNFEPSVPPYPFESK	24,87/0,026
		IPNFEPSVPPYPFESK	37,65/0,0014
		IPNFEPSVPPYPFESK	22,6/0,042
		IPNFEPSVPPYPFESK	36,16/0,0019
		ISDNYSSLLQVSEALGR	57,76/1,40E-05

Supplemental Table 1. List of peptides of PP2CA, BPMs and RGLG1 identified by LC-MS/MS in five independent experiments. Total proteins were extracted from 2w old *A. thaliana* seedlings (*35S_{pro}:3×FLAG-PP2CA*) treated without (mock) or with different mixes and times of 50µM MG-132 and 50µM ABA. Immunoprecipitation (IP) was performed using anti-FLAG and the IP products were analysed by LC-MS/MS.